Introduction to SIP IP Telephony Systems

Technology Basics, Services, Economics, and Installation

Lawrence Harte, David Bowler

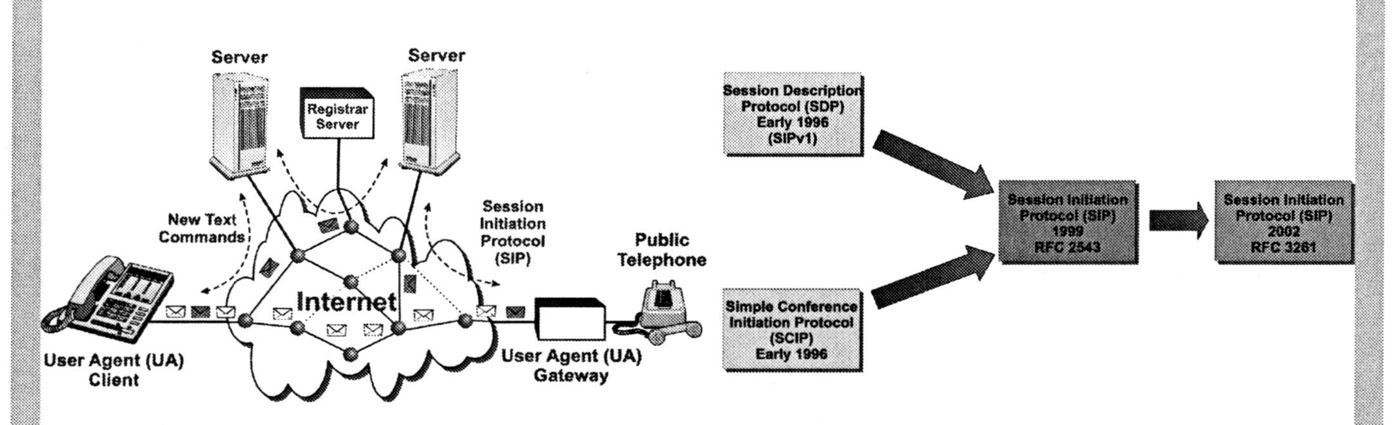

Devices and How they Operate

Standards and their Evolution

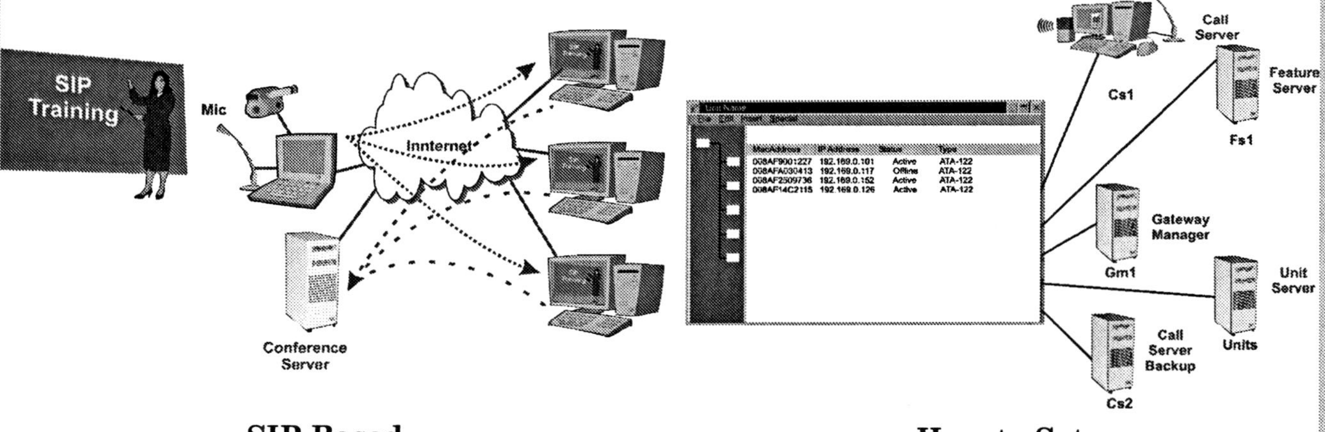

SIP Based Applications

How to Setup SIP Systems

Excerpted From:

ALTHOS Publishing

SIP IP Telephony Basics

ALTHOS Publishing

ISBN: 0-9728053-8-9

ALTHOS electronic books (ebooks) and images are available for use in educational, promotional materials, training programs, and other uses. For more information about using ALTHOS ebooks and images, please contact Karen Bunn at kbunn@Althos.com or (919) 557-2260

Terms of Use

About the Authors

Mr. Harte is the president of Althos, an expert information provider covering the communications industry. He has over 29 years of technology analysis, development, implementation, and business management experience. Mr. Harte has worked for leading companies including Ericsson/General Electric, Audiovox/Toshiba and Westinghouse and consulted for hundreds of other companies. Mr. Harte continually researches, analyzes, and tests new communication technologies, applications, and services. He has authored over 30 books on telecommunications technologies on topics including Wireless Mobile, Data Communications, VoIP, Broadband, Prepaid Services, and Communications Billing. Mr. Harte's holds many degrees and certificates include an Executive MBA from Wake Forest University (1995) and a BSET from the University of the State of New York, (1990).

Mr. Bowler is an independent telecommunications training consultant. He has almost 20 years experience in designing and delivering training in the areas of wireless networks and related technologies, including CDMA, TDMA, GSM and 3G systems. He also has expertise in Wireless Local Loop and microwave radio systems and has designed and delivered a range of training courses on SS7 and other network signaling protocols. Mr. Bowler has worked for a number of telecommunications operators including Cable and Wireless and Mercury Communications and also for Wray Castle a telecommunications training company where he was responsible for the design of training programmes for delivery on a global basis. Mr. Bowler was educated in the United Kingdom and holds a series of specialized maritime electronic engineering certificates.

Table of Contents

Chapter 1

Introduction to SIP

SIP is a standardized technology that allows for the sending of voice, data, and video between communication devices on public (e.g. Internet) or private (e.g. local area networks) data networks.

SIP systems can integrate with or replace traditional private telephone systems. Because SIP systems are based on the same type of data communications protocol that is used for many information systems, SIP systems can be easily be integrated with other systems such as web pages, email, and file transfer programs.

All SIP devices have network addresses. This allows for network management to setup (configure) devices, setup services, and monitor the status of SIP systems.

SIP communication system economics include the cost of equipment, operations, and other hidden costs. SIP has evolved from two different proposed standards into a single text based system. SIP continues to evolve into an advanced multimedia communication system that goes well beyond traditional telephone call processing services.

Basic VoIP Operation

Voice over Internet Protocol (VoIP) is a process of sending voice telephone signals over the Internet or other type of data network. If the telephone signal is in analog form, (voice or fax), the signal is first converted to a digital form. Packet routing information is then added to the digital voice signal so it can be routed through the Internet or data network.

Figure 1.1 shows how an Internet network (public or private) can be used to provide telephone services. In this example, a calling telephone or multimedia capable computer dials a telephone number. These dialed digits are sent to a call server (call processor) that determines that this call must be rout-

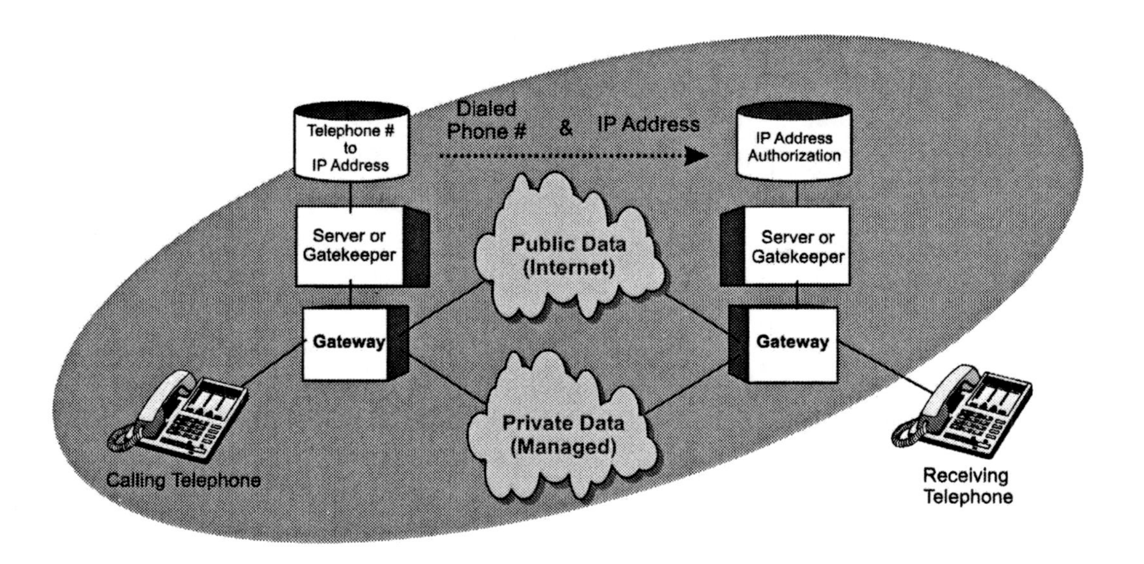

Figure 1.1., IP Telephony Operation

ed to a phone in another area. The call server decodes the dialed digits and determines the destination address (IP address) of the call server that can service the dialed telephone number. The remote call server sends a message to the gateway (data/voice conversion) that alerts the caller of an incoming call (rings the phone or alerts a multimedia computer). When the user answers the call, a message is sent between the call servers and a two-way virtual path (two one-way connections) are created between the devices.

SIP System Overview

SIP is an application layer protocol that uses text format messages to setup, manage, and terminate multimedia communication sessions. SIP is a simplified version of the ITU H.323 packet multimedia system. SIP is defined in Request For Comments (RFC) document 3261.

SIP systems are primarily composed of end-user communication devices (User Agents) and call processing computers (Servers). User agents allow people or devices to have access to SIP systems. Servers receive, initiate, and respond to requests for communication services.

Figure 1.2 shows how a SIP system uses relatively simple text messages to setup and control telephone calls. This diagram shows how a telephone has SIP capability that is controlled by a call server. This SIP based telephone is called a User Agent (UA). The User Agent (UA) is actually a gateway that converts audio (e.g. sound) and control information (e.g. dialed digits) into packets that can be routed through a data network (such as the Internet) to call servers and other User Agents (UAs.) The control packets are sent to and from the call server to request and receive calls. Call servers may communicate with other call servers to setup distant call connections. This diagram shows how a distant call server controls a User Agent (UA) gateway that allows calls to connect from the Internet to another telephone.

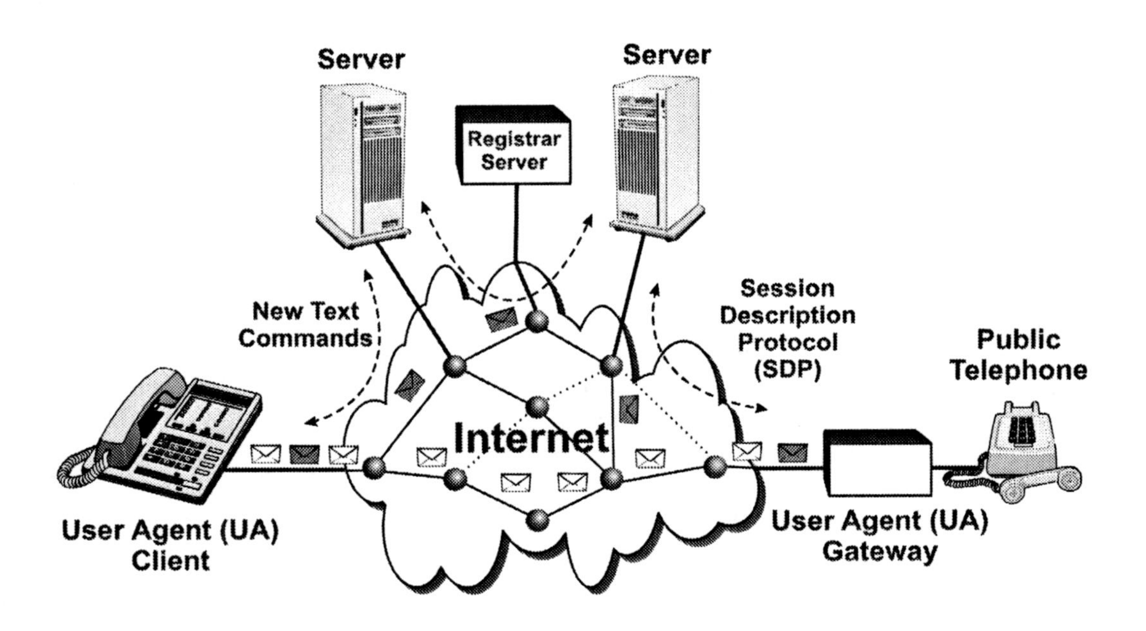

Figure 1.2., SIP System

The SIP protocol allows calls to be established but it does not describe the format of those calls, for example the media to be used. A second protocol, probably the Session Description Protocol (SDP), would be used in conjunction with SIP to describe the media format of the communications session.

Communication Services

Communication services are the processes that transfer information between two or more points. Communication services may involve the transfer of one type of signal or a mix of voice, data, or video signals. When communication services only involve the transport of information, they are called bearer services. When communication services involve additional processing of information during transfer (such as store and forward), they are known as teleservices.

Figure 1.3 shows how multiple forms of media can be sent during an Internet telephone call. This example shows a single broadband connection can simultaneously allow telephone calls (Internet Telephone service), transfer data (such as browsing the web), and allow the display of video. In this scenario, a teacher is presenting a training session to students. Each student can see the professor on their television, they can see the course presentation on the computer monitor, and they can hear the professor by the audio on the computer speakers.

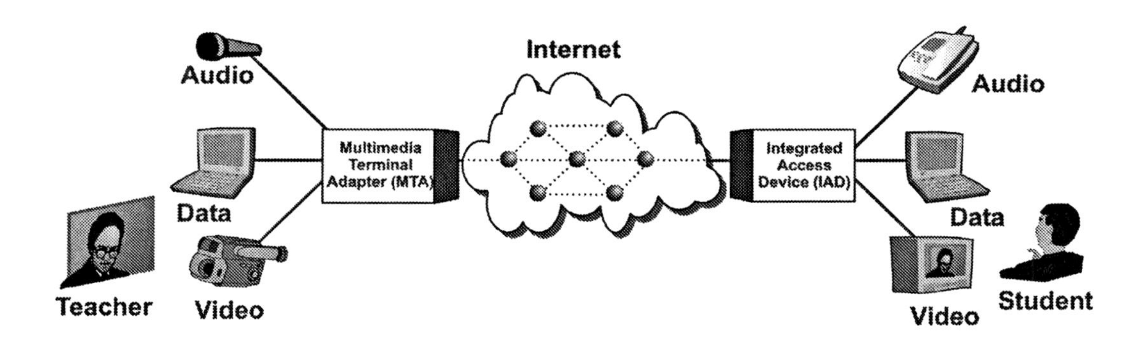

Figure 1.3., Communication Services

Private Telephone Systems

Private telephone systems are independent telephone systems that are owned or leased by a company or individual. Private telephone networks include key telephone systems (KTS), Private [Automatic] Branch Exchange (P[A]BX) and Computer Telephone Integration (CTI).

Figure 1.4 shows the basic types of private telephone systems; key telephone systems (KTS), private branch exchange (PBX), and computer telephony integration (CTI) systems. The most basic private telephone network is a single telephone attached to a business line. Key systems allow each telephone in a business to answer and originate calls on several business lines. A PBX system allows many extensions within a business to call each other and the PSTN. Computer telephony (CT) systems are communication networks that merge computer intelligence with telecommunications devices and technologies.

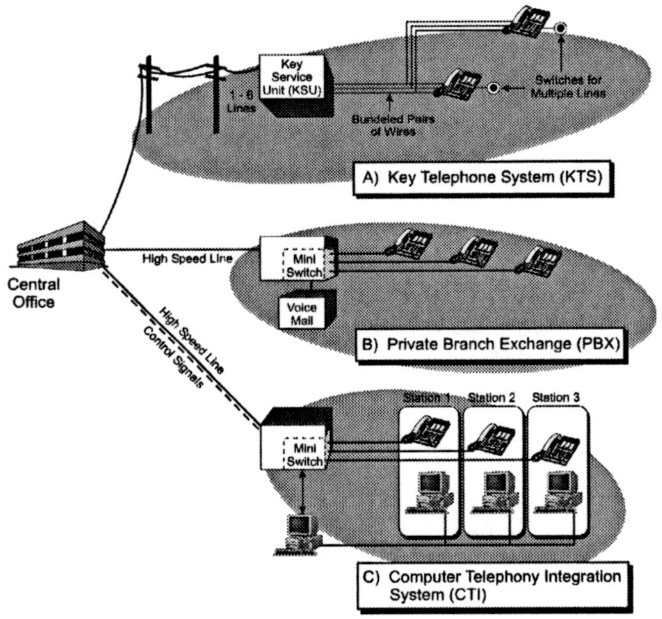

Figure 1.4., Private Telephone Systems

Information Systems

Information systems store, transfer, and process information for specific purposes. Information systems consist of hardware (usually computers) and software (data and applications) that add value to information by; generating, acquiring, storing, transforming, processing, retrieving, utilizing, or making available information via data and telecommunications connections. Examples of information systems include information storage, financial applications, order processing, web e-commerce, and engineering design.

Figure 1.5 shows how an information system is composed of many different types of computer hardware, data networks, and software applications. This example shows that a single company may have several different types of information systems that may be interconnected in a variety of ways. In this example, the company interconnects many of its information systems using a local area data network. This example also shows that this company transfers information to a web host on a temporary link for the e-commerce web site (web host).

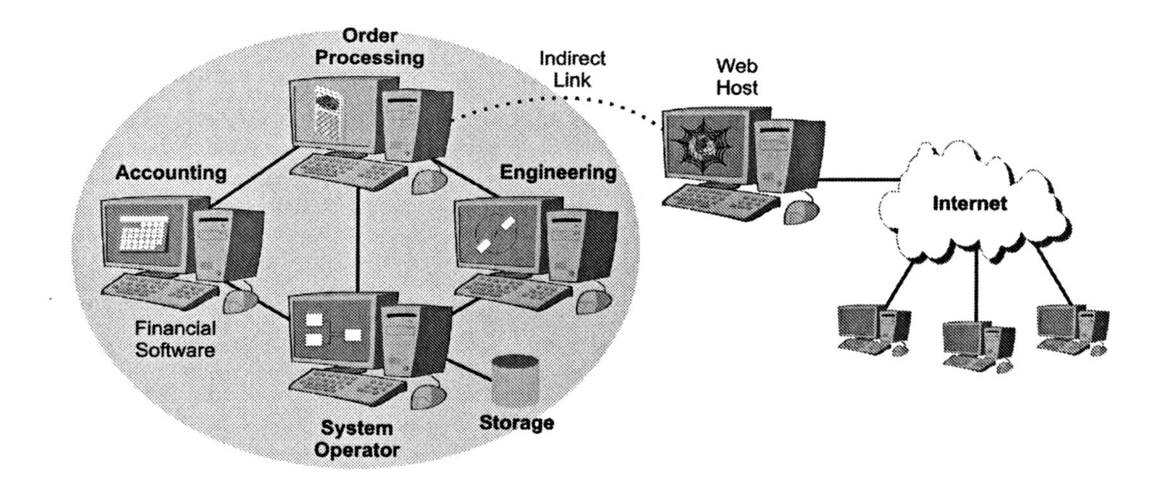

Figure 1.5., Information Systems

Network Management

Network management is the processes of configuring equipment in the network, the setup (provisioning) of services, system maintenance, and repair (diagnostic) processes. Network management systems are commonly composed of a network management server computer and network management software.

Equipment within a network needs to be configured for both hardware (assemblies) and software. Network managers may setup communication services by creating accounts and selecting features for those accounts. Network management software commonly monitors the status of network assemblies to determine if the equipment is operating within normal specifications. When equipment falls outside the desired specifications, network management software may initiate additional tests and reconfigure the equipment to ensure continued network operation.

Figure 1.6 shows how network management is used to configure, setup services, and maintain networks. In this example, a network management server is transferring configuration parameters to a media gateway. The network manager is also setting up (provisioning) services for a call server to allow the call server to route calls through the gateway. After the gateway is setup, this example shows that the network management system will periodically receives messages from the gateway that contain system status information to ensure the network is operating within its desired specifications.

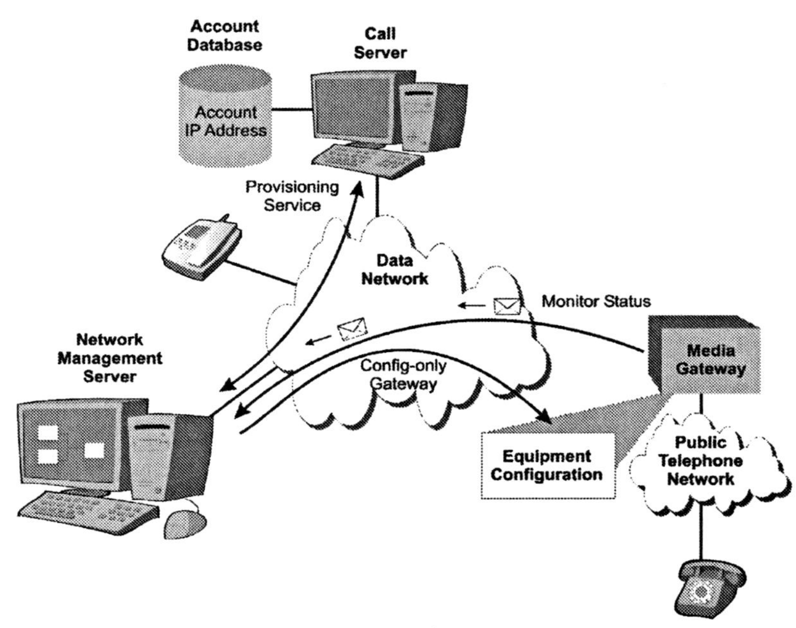

Figure 1.6., Network Management

Communication System Economics

Communication system economics for SIP systems involves fixed (e.g. equipment and software) and operational (e.g. gateway access and data line leasing) costs. Purchasing and setting up a SIP based communication systems is more involved than just buying computer hardware equipment, IP telephones or IP telephone adapters, and software. It includes the setup, operation, and maintenance of multiple types of communication systems.

System costs may also involve the licensing of technology and software, enhancing existing data networks for increased reliability, installation of call servers and information systems, provisioning (activation) of accounts and feature profiles, training for end users and system administrators. All of these costs will increase the average cost per device.

There are a variety of operational costs (usage costs) for SIP systems. These costs may include data transmission line leasing fees, port connection charges, and gateway usage fees. New data transmission lines may need to be leased to interconnect or make redundant paths the communication system to ensure the reliability of the system. The bandwidth of existing data transmission lines may need to be increased to handle the combined voice and data network traffic. For SIP systems that desire to complete calls to public telephone networks where the company does not own media gateways, gateway termination fees of 1 to 2 cents per minute are common.

SIP Evolution

SIP technology evolution involves the development of new features that go beyond the simple call processing features that the original SIP protocol defined. The first generation of SIP technology allowed for basic voice call and conference call capabilities. Because the SIP protocol (the core specification) was designed to setup communication sessions, to modify these sessions as necessary (add or remove media types), and to terminate the sessions, this has allowed SIP to evolve into a powerful multimedia communication system.

SIP was designed to allow for the simple addition of new services and capabilities. The SIP core specification defines the basic media handling capabilities of SIP. As new capabilities are developed, they are introduced as separate extensions (enhancements) to SIP. These extensions define how SIP systems will interoperate with other protocols such as telephone system control (e.g. the Q.931 signaling protocol) and web content delivery (e.g. HTTP). Because access devices have different media handling capabilities (voice, data and video) and the availability of these media formats can change, the SIP core (main program) was designed to be flexible by discovering the capabilities of devices and adapt to its environment for new communications sessions.

The SIP system was designed to easily expand from small systems to very large carriers (scaleable). SIP is scaleable because it defines multiple types of call servers (proxies). These call servers can be tightly coordinated with the user device (called stateful proxies) or these call servers can simply be used to transfer packets for communication sessions (stateless proxies).

Because SIP is independent of underlying data transmission technology, SIP has been selected for use by other communication systems. These systems include 3rd generation wireless, instant messaging services, and cable television systems.

Figure 1.7 shows the evolution of SIP. This diagram shows that SIP evolved from the merging of two similar protocols proposed in 1996, the 1999 version of SIP documented in RFC 2543 has now been replaced by a version defined in RFC 3261.

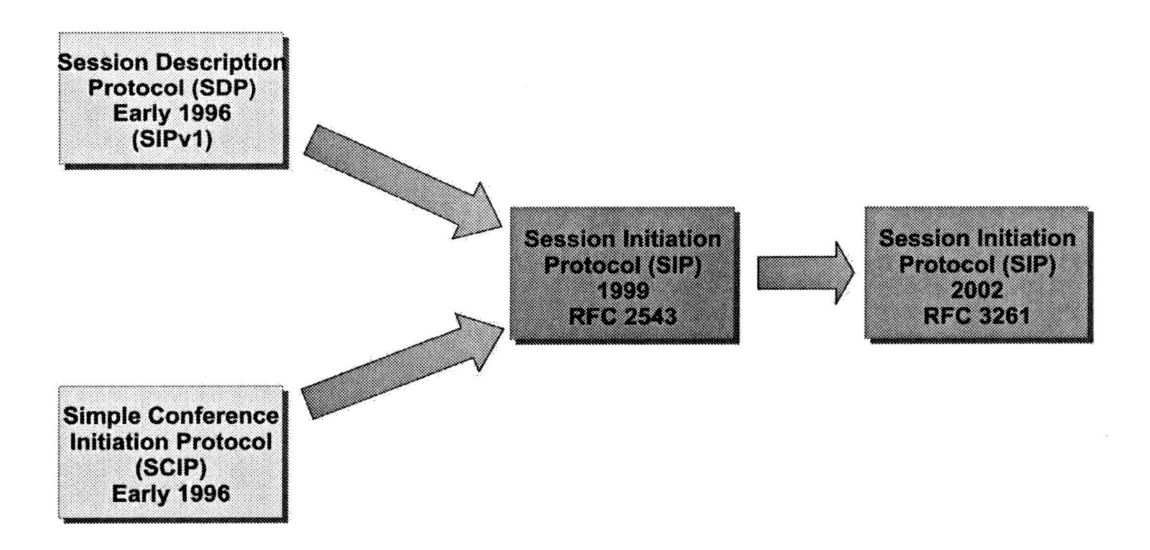

Figure 1.7., Evolution of SIP

Chapter 2

SIP System Operation

SIP communication systems use standard protocols (such as the Internet Protocols) and protocols that were specifically designed for coordinating the SIP system. These protocols are used to control end user devices (called user agents), process call requests (by the means of proxy servers), authorize user (customer) requests for service access (in databases called registrars), track addresses (in location registers), and forward calls (called redirection servers).

Protocols

Protocols are the languages, processes, and procedures that perform functions used to send control messages and coordinate the transfer of data. Protocols define the format, timing, sequence, and error checking used on a network or computing system. While there are several different protocol languages used for Internet telephone service, the underlying processes (setup and disconnection of calls) are fundamentally the same.

Systems can use sets of protocols. There are protocols for call setup, audio compression, call conferencing. Protocols are commonly grouped together into families of protocols to ensure they work together (interoperate) without problems.

Protocols are often enhanced and modified over time as new feature needs and problem areas are identified. As a result, protocols may have different revisions and earlier revisions may have more limited features and capabilities.

Figure 2.1 shows how protocols are used to communicate and control each part of an Internet telephone system and how different protocols can be used in different parts of the network. In this diagram, an Internet telephone is communicating with a public telephone. The Internet telephone creates packets of data. Some of these packets contain voice information and some contain control information (the shaded packets). The Internet telephone protocol defines how these packets are addressed and processed. Control packets are sent to and from a computer (commonly called a gatekeeper, server, or controller) and this computer manages the setup and disconnection of calls and advanced services. The controlling computer communicates with other controlling computers in the network to allow calls to be connected.

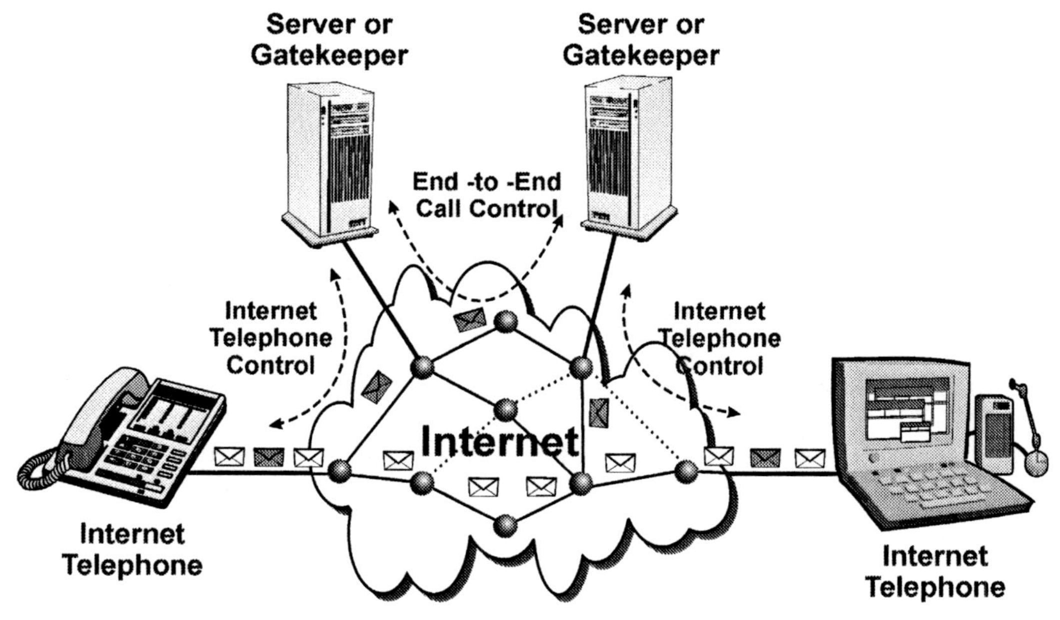

Figure 2.1., Internet Telephone Protocols

If SIP is the network protocol being used then the telephone control and end-to-end control are both performed by the SIP protocol and the servers are known as proxy servers.

The Session Initiation Protocol (SIP)

The Session Initiation protocol (SIP) is a fairly simple protocol that uses text-like messages (text-based) that are transferred on Internet protocol based systems. SIP messages are similar to Hypertext Transfer Protocol (HTTP) messages that are used by Internet web applications.

SIP is an application layer protocol that is used to setup, manage, and terminate multimedia communication sessions. SIP is a simplified version of the ITU H.323 packet multimedia system. SIP is defined in RFC 3261.

SIP is much more simple than the ITU H.323 standard because it has created new commands instead of attempting to adapt commands from established protocols. SIP allows both independent operation (caller to caller directly through the Internet) or SIP can be used by an Internet telephone service provider (ITSP) to manage calls to and from their customers.

End User Devices (User Agents)

End user devices are communication adapters such as telephones, fax machines, or private telephone systems that adapt signals from a telecommunication system to a format (such as audio or visual) to a form that is suitable for an end user.

End user devices in a SIP system are called user agents (UA). The UA is a conversion device that adapts signals from a data network into a format that is suitable for users. Examples of user agents include dedicated IP telephones (hardphones), Analog Telephone Adapters (ATAs), or software (softphones) that operate on a computer that has multimedia (audio) capabilities.

SIP User Agents may perform as service request devices (clients) or may respond to requests from the SIP network to delivery calls (server). When the User Agent operates as a requestor of service, it is called a user agent client (UAC) and when it is responding to requests for service, it is called a user agent server (UAS).

Figure 2.2 shows how a SIP system uses relatively simple text messages to setup and control telephone calls. This diagram shows a user agent (UA) SIP telephone that is controlled by a call server. The user agent (UA) telephone is actually a gateway that converts audio and control information into packets. The control packets are sent to and from the server to request and receive calls. Servers may communicate with other servers to setup distant call connections. This diagram shows how a distant server controls a user agent (UA) gateway that allows calls to connect from the Internet to a public telephone.

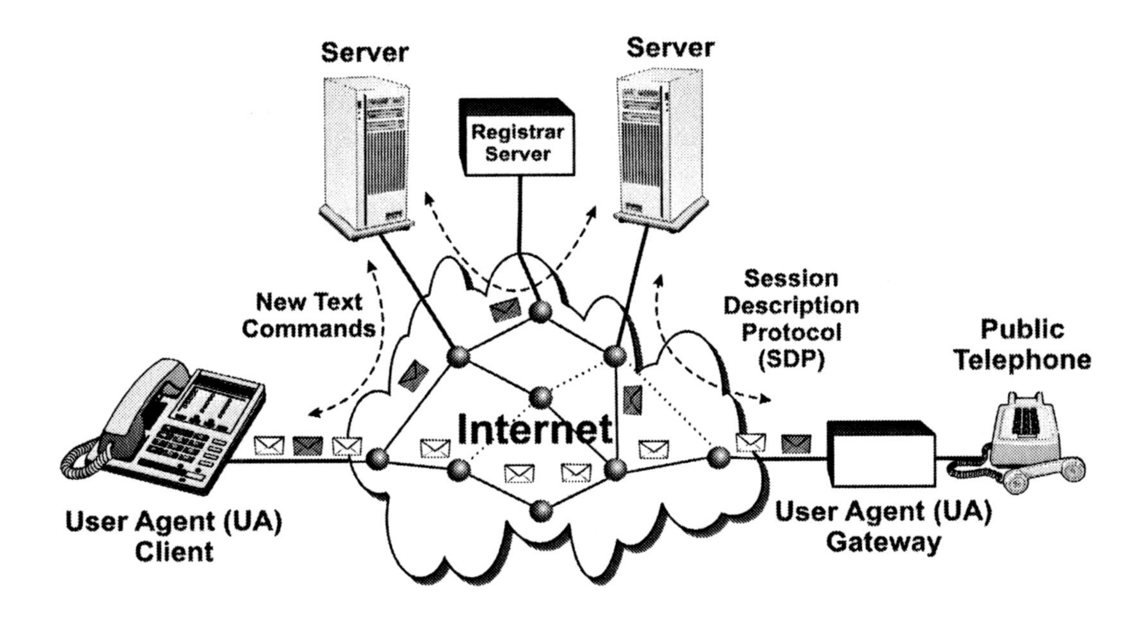

Figure 2.2., SIP System Overview

Proxy Servers (Call Processors)

Proxy servers are computing devices (typically a server) that interface between data processing devices (e.g. computers) and other devices within a communications network. These devices may be located on the same local area network or an external network (e.g. the Internet). A proxy server usually has access to at least two communication interfaces. One interface communicates with a device requesting services (e.g. a client) and a device that is being requested for a service (the server).

The SIP proxy server performs call processing tasks on the behalf (as a proxy) of another device. Proxy servers receive requests (such as invite requests to start a communication session) and perform processes to assist in the establishment of the communication system. This process may only involve the forwarding of requests or it may involve the changing (processing) of information as it passes through the proxy server. SIP proxies may modify or create new SIP messages based on the requirements of the communication session and the setup of the services authorized for the proxy.

Proxy servers may forward call requests to more than one user agent (called forking proxy). This forwarding may be sequential (such as searching through a list of numbers or addresses) or they may be in parallel (such as ringing several extensions at the same time).

Figure 2.3 shows how a SIP proxy server is responsible for establishing communication connections between devices within a specific domain. In this example, when the proxy server receives a request message from a User Agent (UA) inviting another party to join a session, the proxy server forwards this invitation onwards (it acts as a proxy) to the designated User Agent. If the designated User Agent is unavailable, the proxy server may direct the connection request to the second User Agent or the connection request may be forwarded via one (or more) proxy servers.

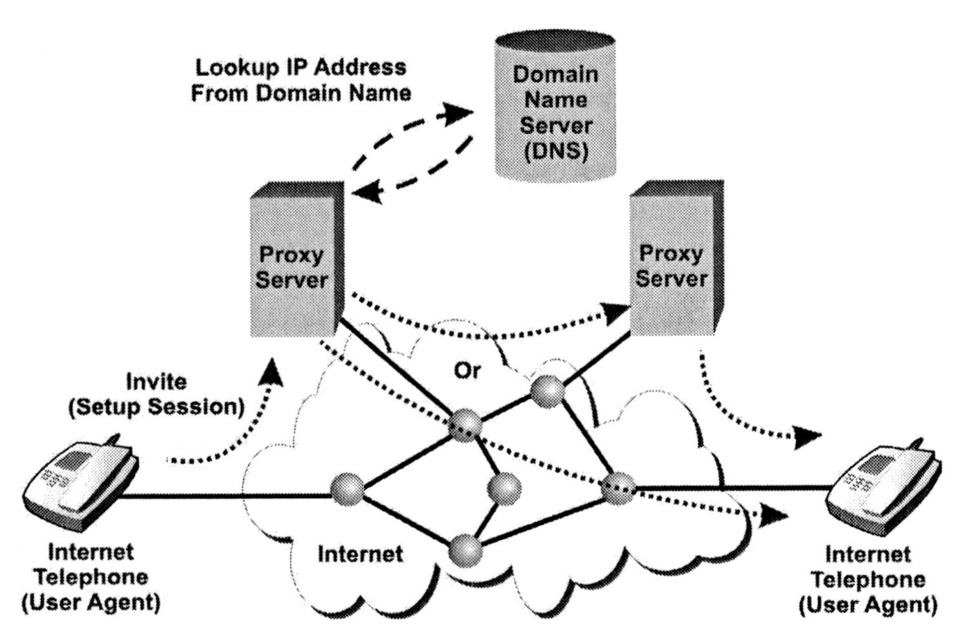

Figure 2.3., SIP Proxy Server Operation

The proxy servers can communicate with elements other than User Agent and other proxies, for example to determine the IP address of another server a proxy may need to call on the services of the Domain Name System (DNS) to resolve a domain name into a IP address. Also the proxies are able to interact with an element known as a Location Service that is used to identify the present location of a particular user.

Registrars (Customer Status)

A registrar is a server that accepts registrations from users and places these registrations, (which are essentially location information), in a database known as a Location Service. The process of registration associates a user with a particular location, (IP address); this association is known as a 'binding' in SIP. When there is an incoming session for a user within a domain, the proxy server will interrogate the Location Server to determine the route for the signaling messages.

Although the proxy and the registrar may be collocated, for the purposes of the SIP protocol they are logically two distinct elements.

Figure 2.4 shows a SIP Registrar that is used to gather and store registration data into a database to provide Location Service. In this example, an Internet telephone user agent is connected to the Internet. The Internet telephone had been previously programmed to register with a specific registrar service on the detection of new service (sensing an Internet connection). When the Internet telephone sends its registration information to the registrar (with its dynamically assigned IP address), the registrar updates the Location Service database to allow calls to be routed correctly for SIP communication sessions.

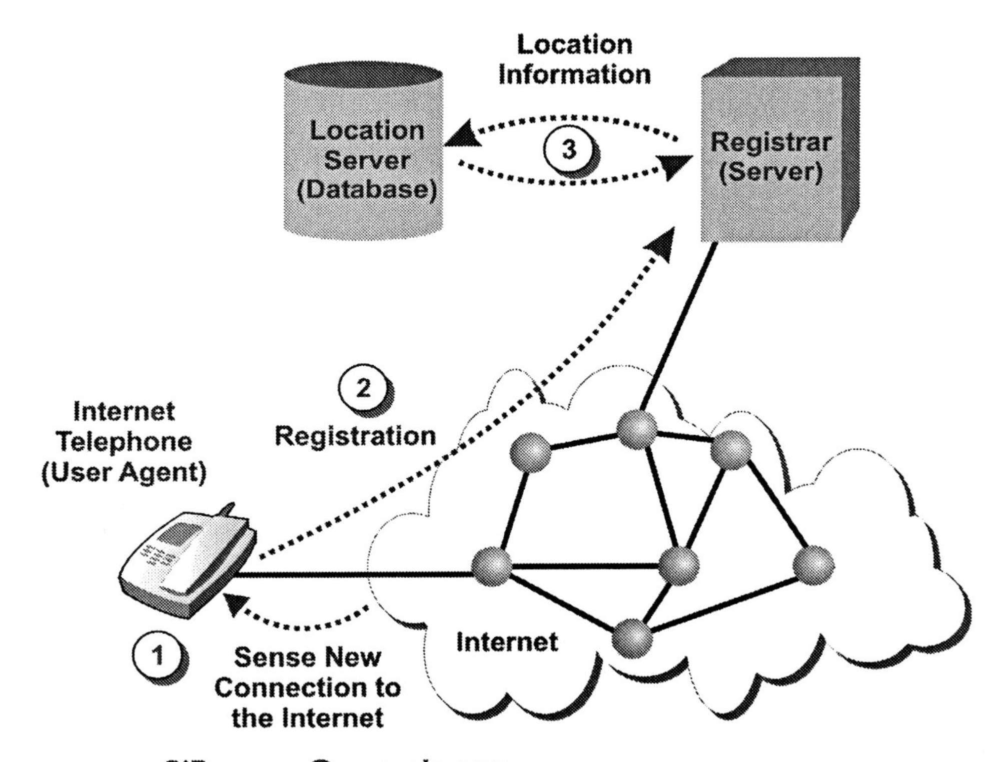

Figure 2.4., SIP Registrar Operation

There are a number of ways in which a User Agent may discover a Registrar. The first of these is to pre-configure the address of the Registrar in the UA, the second method is to use the host portion of the users SIP address to route the registration messages, so if a user had a SIP address of SIP:bill@example.com the UA would route registrations to SIP: example.com. The final method that a UA may use to discover a Registrar is achieved using a multicast address that is sent out by the UA and would be responded to by Registrars that received this multicast.

Location Servers (Address Tracking)

Location servers provide information regarding the location of resources that are located within a network (such as the Internet). Location servers are typically databases that maintain a binding (mapping) for each registered user. This binding maps the address of the user to one or more addresses at which the user can be currently reached.

Location servers regularly exchange information with servers in other administrative domains. Location servers may use the Telephony Routing over Internet protocol (TRIP) to allow the discovery of available devices and gateways.

By means of the registration function of SIP and location services, SIP systems can provide a range of user mobility options. This permits a user to connect to service at any appropriate terminal on any sub-network (such as the Internet). When the user connects their device, it registers with a registrar with updates the information contained in the location server.

Figure 2.5 shows how a location server may be used to assist in the routing of calls in a SIP system. This example shows that a user agent has requested the initiation of a communication session with another user. When the invite request is received by the proxy server, the proxy server sends a location query to the location server to requests the current IP address of SIP:susan@example.com. When the location server returns the last registered address of SIP:susan@example.com, the proxy server forwards the invite request to Susan's User Agent.

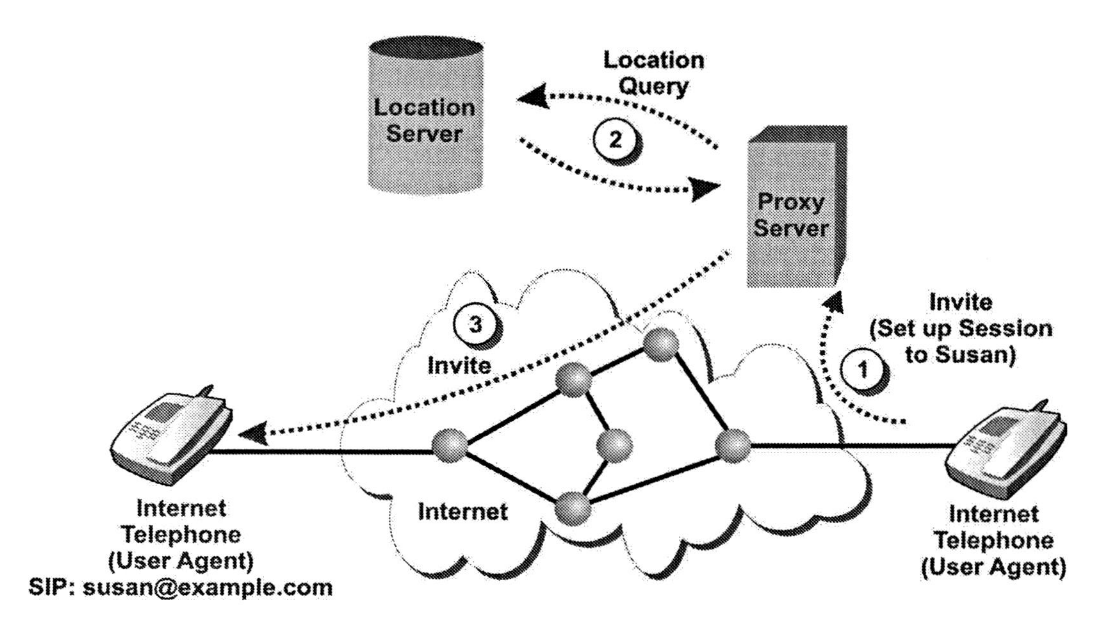

Figure 2.5., SIP Location Server Operation

Redirection Servers (Call Forwarding)

A redirection server assists in the establishment of communication sessions by providing alternative locations where the designated recipient can be found. The redirect server does not initiate any action. It only provides information back to the requesting device as to the potential locations of the designated recipient.

Redirect servers can be deployed in the 'heart' of a network to reduce the signaling processing load on proxy servers by pushing the responsibility for routing to servers at the edge of the network. A redirect server will have access to a function such as the Location Service and when it receives a request it will dip into the Location Service database and push the request back to the client that originated it. The redirect server will supply the client with routing information related to the target of the original request.

Figure 2.6 shows the operation of a SIP redirect server in a SIP network. In this example, a user agent's proxy server receives a request from a client that specifies the SIP address of a recipient's address that is located on a redirect server (such as a company iPBX system). The proxy server forwards this invite request to the redirect server. The redirect server looks at the SIP address in the request and fetches from a database the current location information for this address (an IP address). The redirect server returns a response address to the originating proxy server and this allows the proxy server to redirect its invite request to the required destination.

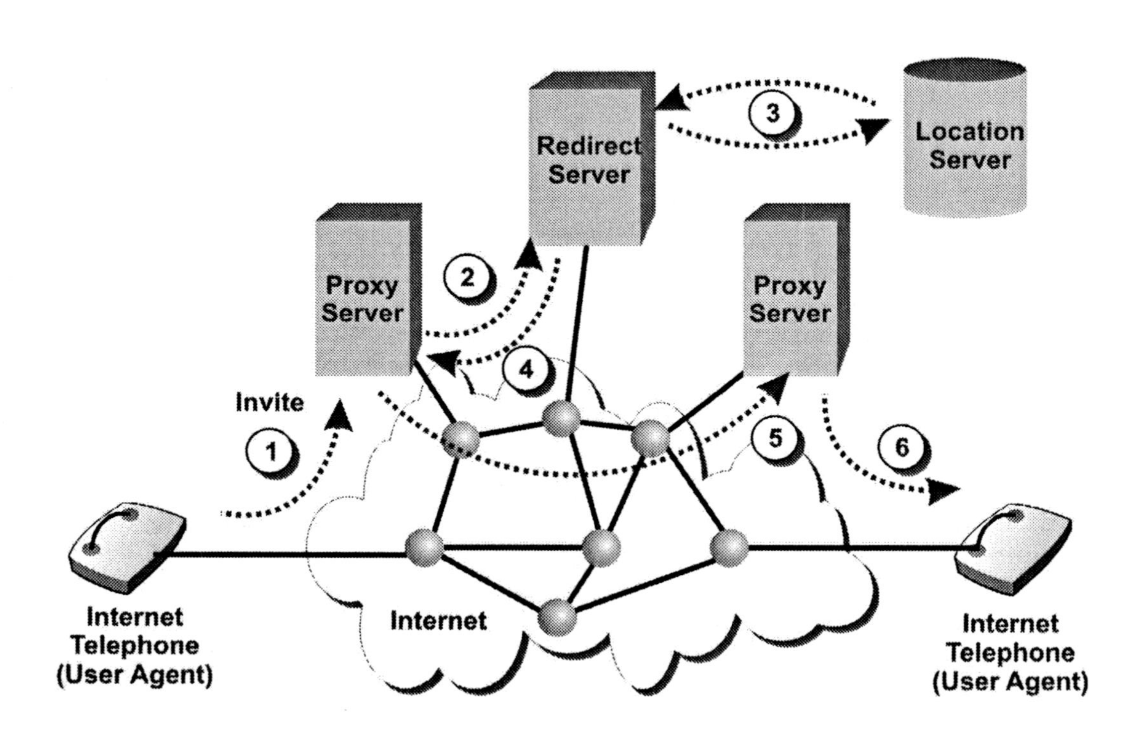

Figure 2.6., SIP Redirect Server

Chapter 3

Basic SIP Communication Services

SIP communication services are the setup, management, and disconnection of communication sessions between two or more users of information. SIP communication services permit the independent or combined transfer of voice, data, and video signals.

To provide SIP communication services through a communication network, the session initiation protocol (SIP) was developed. The SIP protocol is intentionally quite simple in it's operation, yet capable of providing a range of services including basic voice telephony but also more advanced call features such as user mobility, supplemental call processing features such as call hold and call forwarding, and integrated services such as click to dial.

SIP is intended to support a full range of multimedia sessions between users and therefore once a SIP call is established, the same connection that is used for voice service can be used to transfer other information such as images, sounds, or a combination of any media that can be transferred through the communication network (such as the Internet).

Voice Service

Voice service is a type of communication service where two or more people can transfer information in the voice frequency band (not necessarily voice signals) through a communication network. SIP based voice service involves the setup of communication sessions between two (or more) users that allows for the real time (or near real time) transfer of voice type signals between users. In a SIP system Voice services is established by specific types of call processing steps.

The quality of voice services provided by SIP systems can vary dependent on a variety of factors including the amount of data transmission channel quality, and compression method used. The data transmission channel quality can vary based on the transmission delay and the amount and type of errors. The compression methods supported by SIP range from standard 64 kbps pulse coded modulation (PCM) voice to 8 kbps (highly compressed) G.729 speech coding.

The speech coding method is negotiated on call setup. The standard 64 kbps speech coder can provide for both voice and data modem (e.g. fax) transmission. The highly compressed G.729 speech coder is can not used to transfer fax signals or dual tone multi-frequency (DTMF) tones.

Figure 3.1 illustrates a simplified sequence for a SIP call between two users. In this example Larry is going to place a call to Susan over the Internet, Larry has a SIP telephone whilst Susan is using a softphone, (a piece of software running on a multimedia PC). Larry and Susan belong to different domains and each domain contains a SIP proxy server that manages that domain. The call would be initiated by Larry dialing a number for Susan, or alternatively by selecting an entry from an address book or even a link on web page, (called click to dial).

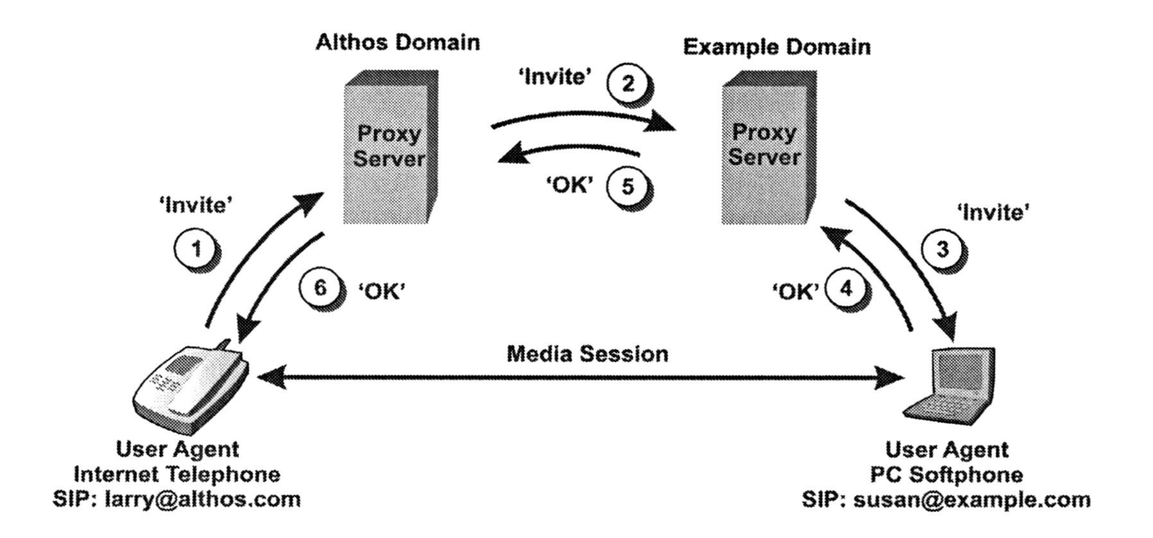

Figure 3.1., SIP Telephone Call

When the call is initiated Larry's phone sends a message to the proxy server for his domain, this proxy server will send a message on to the proxy server in Susan's domain. The Althos.com server may use a form of Domain Name Server (DNS) lookup to obtain the address of the example.com proxy server. If necessary the example.com proxy consults a database known as a location server to identify the current address being used by Susan and forwards the message on to Susan's User Agent, which then generates a message in response that is sent back via the two proxies.

Larry's User Agent will respond with an acknowledgement, but note that this acknowledgement is not necessarily sent via the proxy servers. A two-way media session is now established between the users. When the session is complete the 'connection' can be released by means of a simple handshake between the two telephones.

It is important to note in the call example of Figure 3.1 that the SIP protocol does not define the media format to be used during the call. Instead the SIP messages will convey information from another protocol to define the media to be used during the communications session. In most cases this additional protocol is likely to be the Session Description Protocol (SDP).

Mobility Management (via Registration)

Mobility management is the processes of continually tracking the location of telephones or devices that are connected to a communication system. Mobility management typically involves regularly registering telephones or communication access devices. Mobile telephones typically automatically register when they are first powered on or attached to the communication systems. Some devices may also register and when they are powered off or detached from the system.

The SIP protocol supports user mobility, by allowing a user to both initiate and receive sessions on different terminals within a domain, also a user is able to participate in session on terminals outside of their home domain (such as being attached anywhere to the Internet). Servers known as Registrars provide mobility in a SIP system. A Registrar has an associated database, known as the Location Service, which is used to bind a user's SIP address to a current location (IP address). A SIP User Agent can be setup to register with the SIP Registrar when it is first connected to a data network. This allows the Registrar to maintain the latest address (IP Address) where the User Agent is located.

Figure 3.2 shows an example how a SIP system can allow users to attach their devices anywhere within the data network and maintain their ability to make and receive calls. In this example, a User Agent is registering with its Registrar. The User Agent at which Susan is currently located sends a registration message to the Registrar and the Registrar sends this data to be stored in the Location Service database. This creates what is known as a binding between Susan's SIP address and the User Agent she is currently utilizing. When another user, in this example Larry, attempts to establish a session with Susan, the proxy server for Susan's domain will make a query to the Location Service that will return the binding information. This allows the invitation request to be routed from the proxy to the User Agent for Susan.

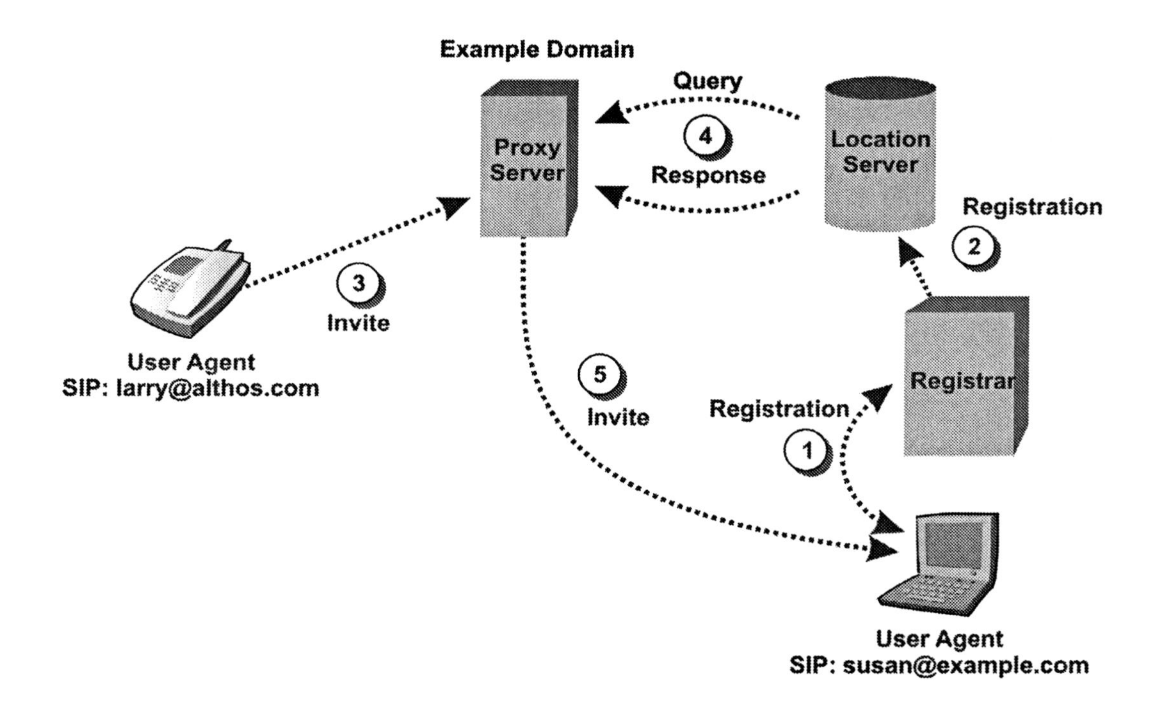

Figure 3.2., SIP Mobility Management

In addition to supporting the basic facility of establishing a call between two, or more users, SIP supports a range features that most users will familiar with from their existing telephony systems. These features include call hold, call forwarding, three-way calling and automatic redial.

Call Hold

Call hold is a feature that allows a user to temporarily hold an incoming call, typically to use other features such as transfer or to originate a 3rd party call. During the call hold period, the caller may hear silence or music depending on the network or telephone feature.

Figure 3.3 shows how a SIP call can be temporarily placed on hold so the call can stay connected without the user having to continue conversation with the caller. During call hold, the media streams in both directions are normally halted. However, SIP can redirect a communication session to provide music on hold. In this case, the party that is placed in the hold condition will be sent a media stream that contains music. A different media server might provide the music.

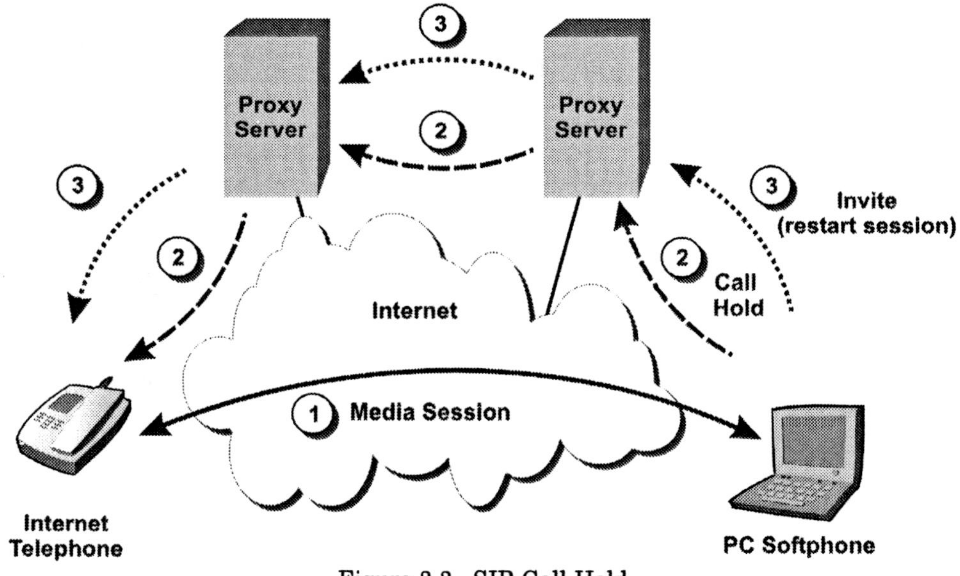

Figure 3.3., SIP Call Hold

Call Forwarding

Call forwarding is a call processing feature that allows a user to have telephones calls automatically redirected to another telephone number or device (such as a voicemail system). There can be conditional or unconditional reasons for call forwarding. If the user selects that all calls are forwarded to another telephone device (such as a telephone number or voice mailbox), this is unconditional call forwarding. Conditional reasons for call forwarding include if the user is busy, does not answer or is not reachable (such as when a mobile phone is out of service area).

The support of call forwarding in any system requires that at least one network element is aware of the user's call handling preferences. For example, under what conditions, if any, should forwarding be triggered and when it is triggered where is the call to be forwarded to? In a SIP system, the user's proxy server contains the call forwarding parameters.

Figure 3.4 illustrates how a SIP system can provide conditional call-forwarding services. In this example, Susan has called Larry and the proxy server invites Larry's User Agent to join the session. Assuming Larry has set the condition 'call forward on no answer' and he is not available to answer this call, the User Agent would ring, or give some other form of alert, for a short (configurable) period of time. When this time expires, the proxy server for the call will forward (by sending an invite request) to a predefined number. In our example, Larry has requested that the call be forwarded to a second SIP address. Call forwarding on busy is very similar to this example except that Larry's User Agent would return a busy indication to the proxy when it received the session invitation.

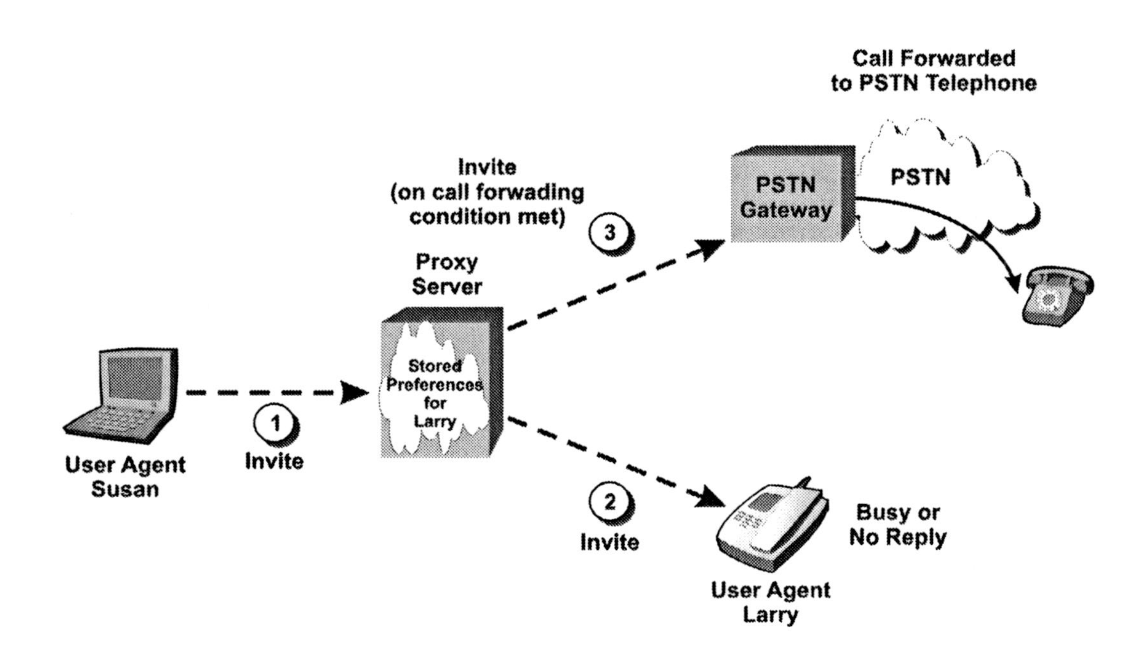

Figure 3.4., SIP Conditional Call Forwarding

Figure 3.5 shows that to implement unconditional call forwarding, a proxy server simply forwards the invitation request to the diverted address previously specified. In this example, Susan has called Larry and the proxy server invites Larry's User Agent to join the session and Larry has set the unconditional call forwarding to a second SIP address. Because the call forwarding has been setup as unconditional call forwarding, the proxy server immediately sends (forwards) the invite request to the designated recipient (forwarded number).

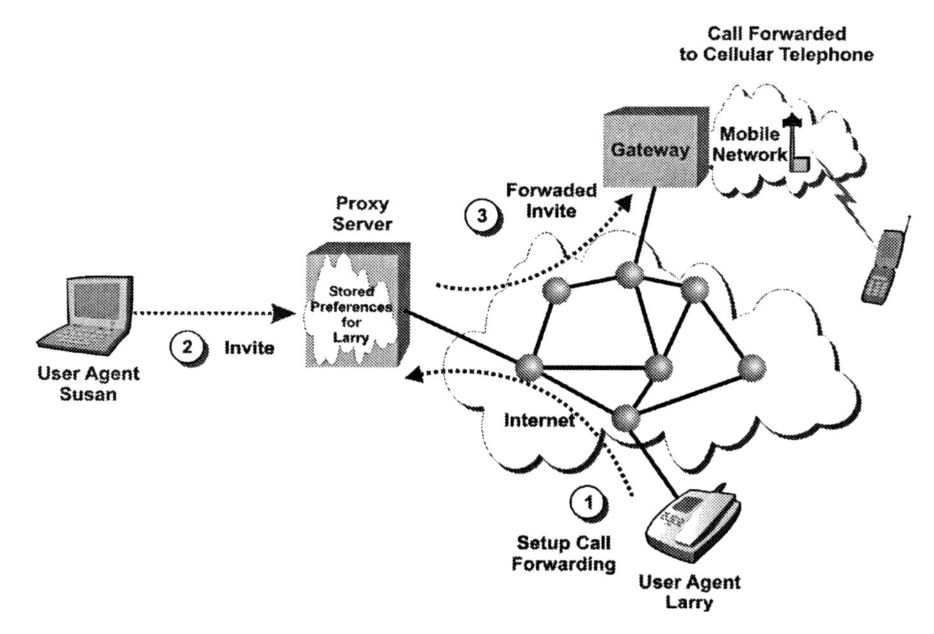

Figure 3.5., SIP Unconditional Call Forwarding

Click To Dial

Click to Dial is a SIP service that allows a user who is viewing a web page to click a link on that web page to initiate a voice over Internet call. The link contains an embedded address (URL or IP address) that connects to a call server along with the necessary software (such as SIP) that allows for the setup and connection of the call. Click to dial service is similar in concept to the 'mailto:' link that can launch a user's email software when selected.

SIP integrates well with web pages to provide click to dial service because SIP is a text-based protocol that interacts with a web browser in similar processes as well established protocols such as Hypertext Transfer Protocol (HTTP) and Simple Mail Transfer Protocol (SMTP). Like these other protocols SIP employs a request-response transaction between entities, such as a User Agent (UA) and Proxy Server.

Figure 3.6 illustrates how click to dial SIP service operates. In this example, a user Susan is viewing a web page that contains a "Click to Dial" button. This button is linked to Larry's SIP address. When Susan clicks on this click to dial the link, it connects her User Agent (also on Susan's PC) to the server at the SIP address provided by the Click to Dial button. The User Agent can then establishes the call by means of the normal SIP call processing (signaling) sequences.

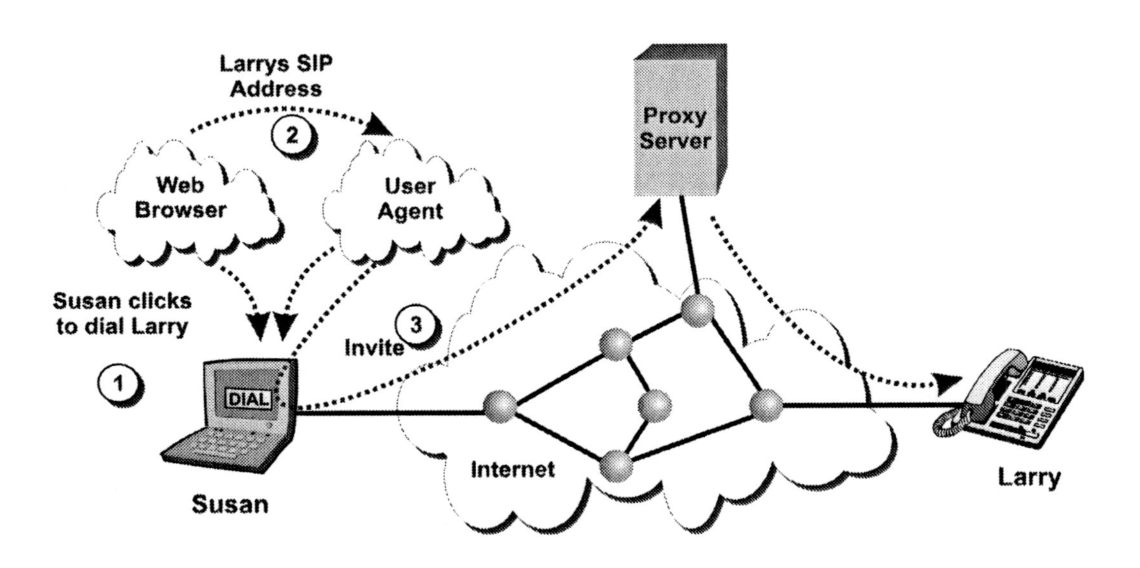

Figure 3.6., Click to Dial Service

Chapter 4

Integrating SIP with Existing Phone Systems

The options for setting up voice over data network service range from the direct connection of existing computers through the Internet to the replacement of PBX telephones and equipment with new IP telephone equipment and software.

Direct Connect

Direct connection is the communication process where two communication devices (such as IP telephones) directly connect to each other through a data network such as the Internet (with or without the assistance of an call server).

Because multimedia computers usually have speakers and microphones, and they can support Internet telephone software, it is relatively easy to setup multimedia computers to directly connect to each other via the Internet for Internet telephone service. The biggest challenge of direct connection is discovering the Internet address of each user (IP telephone or computer). This can be further complicated by the use of private Internet addresses (temporary addresses assigned by a router) and the blocking effect of firewalls.

An easy way to directly connect through the Internet is to use an instant messaging (IM) systems. IM systems were initially setup to allow users to chat with each other by sending and receiving text messages via the Internet. Computers that are connected to the Internet can send data packets to any other computer that is connected to the Internet. Unfortunately, public Internet addresses can change so a solution was an IM computer server that could be used as an address book for users (IM clients) that want to directly communicate with each other. Each instant messaging user will sign on (register) and sign off (de-register) with the IM server each time they want to be able to send and receive messages to other IM members.

Because instant messaging systems keep track of Internet addresses, it is relatively easy for instant messaging systems to offer voice services in addition to text messaging. The user only has to install the instant messaging voice software (Internet telephone software) that can retrieve the Internet address of other users they want to talk to via the instant messaging server. Once the caller has obtained the instant messaging address, the two computers can directly communicate without the need of the instant messaging system to process the call.

Figure 4.1 shows how a company can directly connect VOIP telephony users through the addition of VoIP Software to existing multimedia computers. In this example, Bob in the New York office is calling to Susan in Toyko. Both Bob and Susan have VoIP software (softphone) installed on their computers and they know each others IP address or screen name. Bob initiates a call to Susan by sending a call request (invite) to Susan through they company's data network (public Internet or private data network). When Susan's computer receives the call invite request, it alerts Susan with a screen and/or audio alert (Ring) message. If Susan accepts the call request, the two computers will automatically negotiate the audio parameters (and other parameters such as video). Two call paths are then setup; one in each direction.

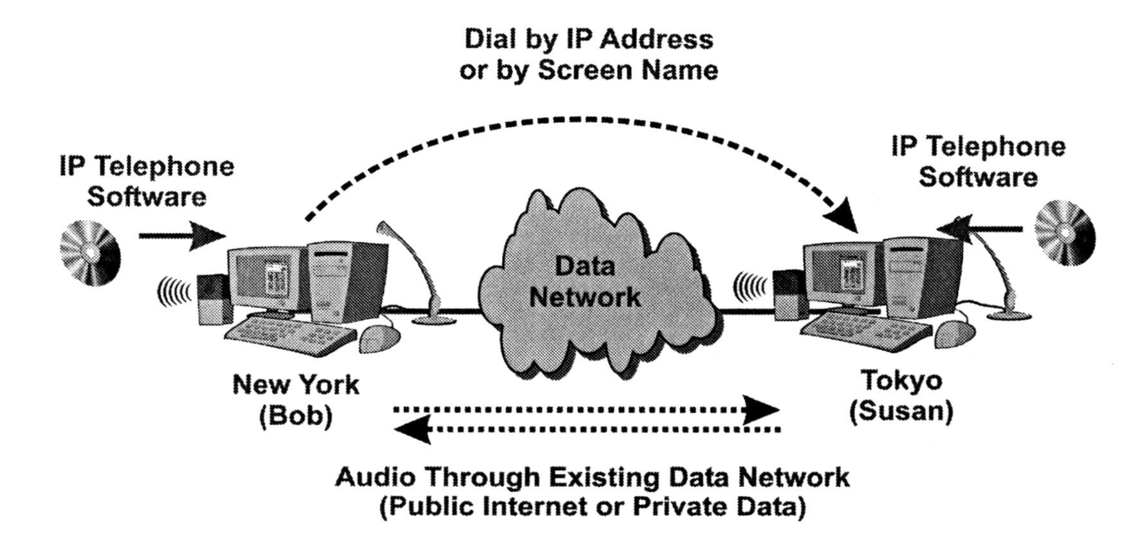

Figure 4.1., Internet Telephone Direct Connection using Instant Messaging Screen Names

Adding VoIP Lines to a Data Network

It is possible to simply add new telephones lines for your company without performing any changes to the existing telephone line devices or connections. Adding new lines is a good way to try voice over data telephone service to determine if it is the right choice without having to change existing systems and services. If the voice over data telephone line service is acceptable, you can use the new telephone lines to gradually replace the standard telephone lines.

Figure 4.2 shows how a PBX system can be upgraded to use voice over IP (VoIP) telephone service without any significant changes to the PBX system. This example shows that a VoIP gateway is used to create addition lines for the PBX system. Because the VoIP gateway can produce standard telephone signals (standard telephone or T1/E1 lines), it can be directly connected to PBX system line cards. The PBX system administrator simply configures the PBX system to use the added lines through the same process that would be used when installing new telephone lines. The gateway is controlled to provide telephone services by a call server from a ITSP, IP Centrex or even another iPBX system.

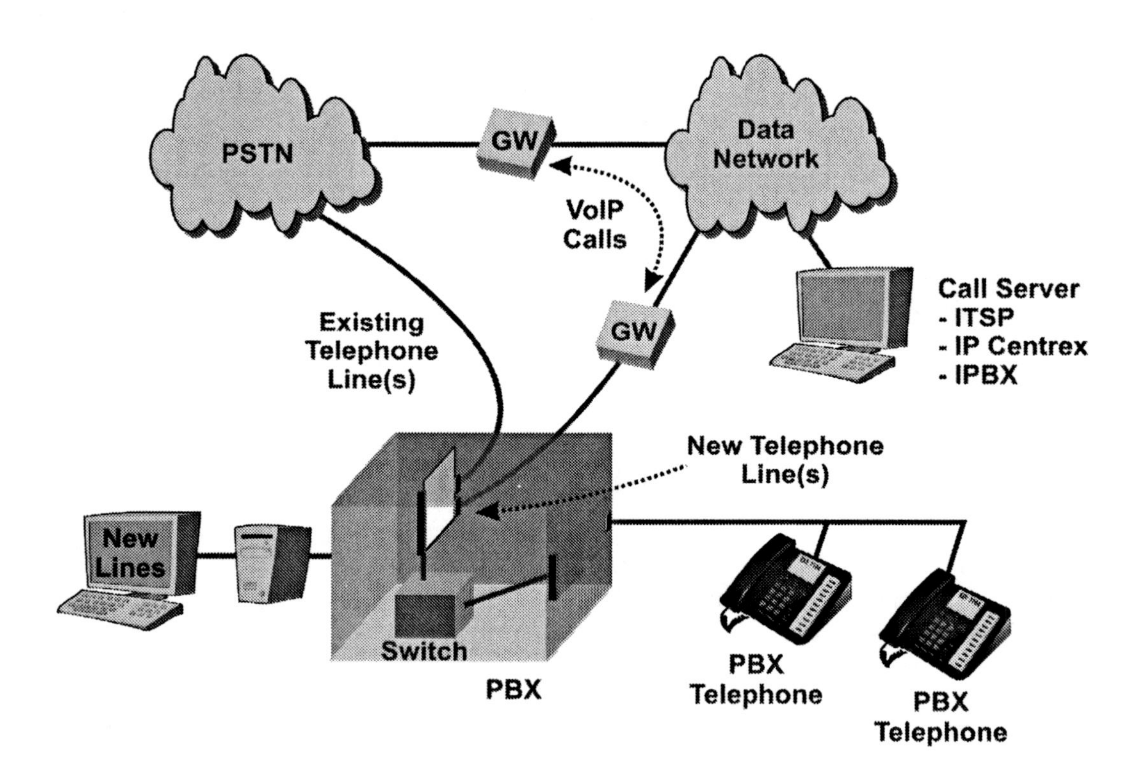

Figure 4.2., Adding Telephone Lines to Company Data Networks

Integrated Service

It may be possible to upgrade standard telephone systems (such as a PBX) with voice over data network systems. Some PBX systems include the option to add processing systems and software that allow the PBX system to setup calls on standard telephone lines or on voice over data (VoIP) lines. These integrated solutions usually allow the PBX system operator decide (assign) which calls should be routed through the Internet and which should go through the standard telephone network.

Figure 4.3 shows how it may be possible to integrate (or upgrade) existing private telephone systems (e.g. PBX) with voice over data (VoIP) service. In this example, a PBX system is upgraded by the addition of a new line module along with software that allows for control of the VoIP line card and its advanced features. In this example, the PBX system continues to use the standard PBX telephones. This example shows that calls to or from PBX

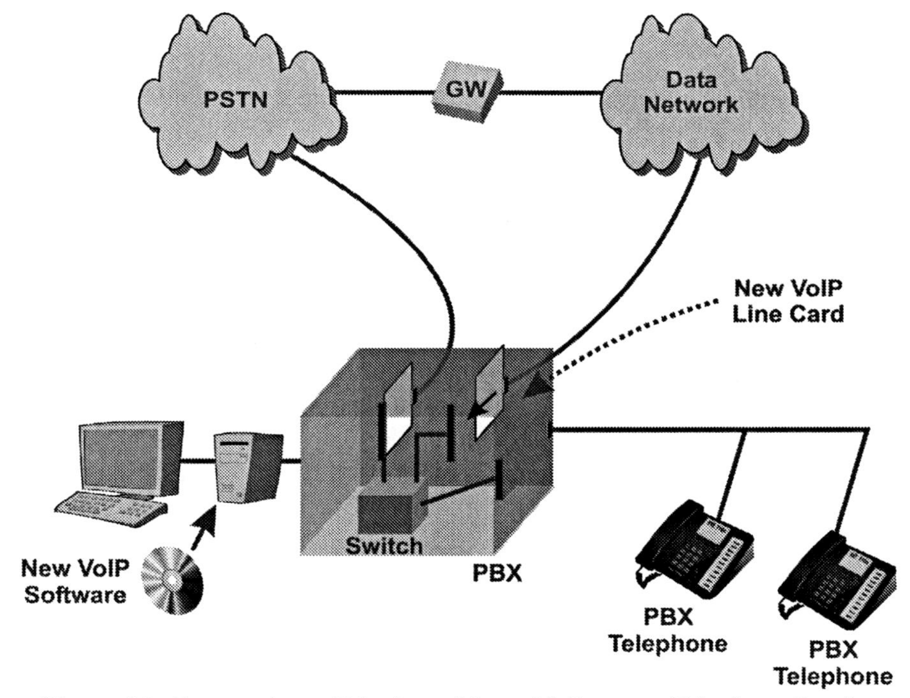

Figure 4.3., Integrating a Telephone Line with Internet Telephone Service

handsets can be connected added to either the public telephone lines or through the VoIP line card to a data network. This upgrade allows the PBX system to connect through data networks to other VoIP gateways that convert calls from VoIP signals to signals that can be connected to the public telephone network.

Replace a System VoIP Lines

Replacing a system involves using Internet telephone service to replace existing telephone lines and systems. The key benefit of replacing a line with Internet telephone service is the elimination of the cost of a standard telephone lines and the ability to use a single low cost data network for both voice and data communications.

One of the key challenges of replacing lines or systems with VoIP telephone service is the telephone number. Few voice over data network service providers (IP Centrex or ITSP) have the ability to transfer your existing telephone number to Internet telephone service (covered in the number portability section). This means if you want to keep your existing telephone numbers (in addition to a new telephone number assigned to you by IP Centrex or ITSP company), you will have to use remote call forwarding. Eventually, many IP telephone service providers will be able to offer number portability and this will allow you to keep your telephone number when replacing a telephone line with Internet telephone service.

Figure 4.4 shows how a company can replace its PBX system with voice over data (VoIP) communication lines and a call server. This diagram shows a private telephone (PBX) system that has been replaced (a "forklift upgrade") by a call server and IP telephones. The IP telephones communicate with the call server that sets up, manages, and disconnects telephone calls through the data network. When the calls require connection to a public telephone line (for an off-data net call), the call server connects the call through a voice gateway (or one of several voice gateways) to connect to the public telephone network.

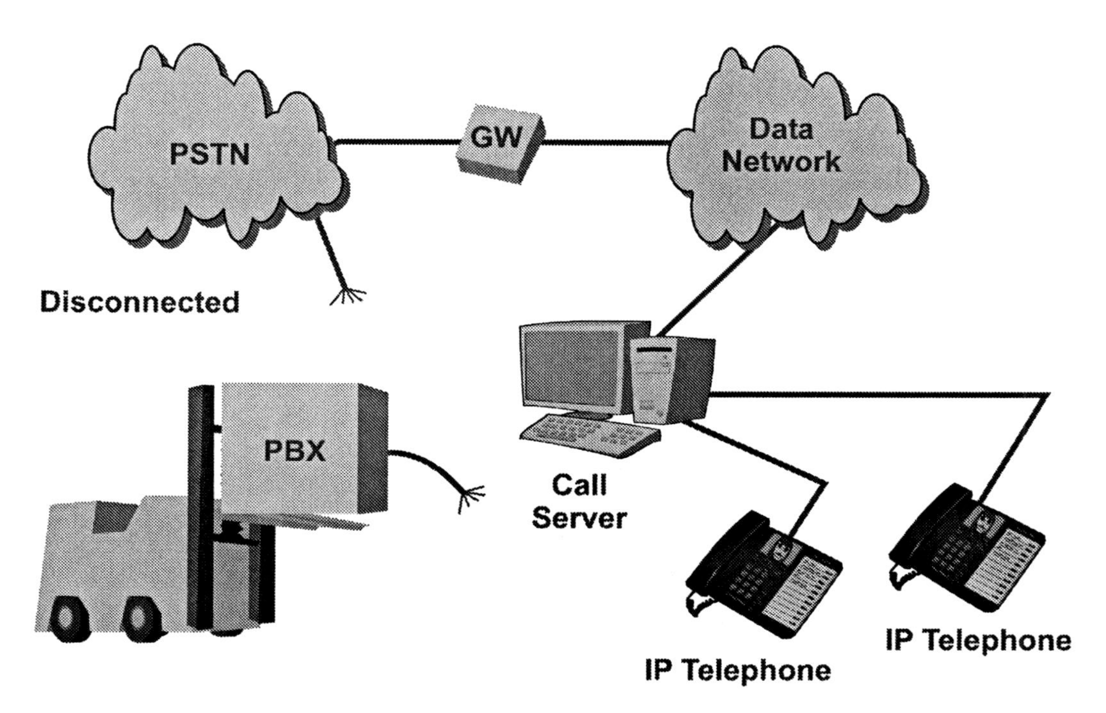

Figure 4.4., Replace Telephone Lines and Systems with Voice over Data Systems

Chapter 5

Tying SIP to Other Information Systems

SIP systems can be easily integrated with multiple types of information systems and other communication networks to produce advanced communication services. SIP systems can be interconnected to other information systems through the use of specialized application servers (AS) and packet data networks. Application servers are computers and associated software that are connected to a communication network to provide information services (applications) for clients (users). Application servers are usually optimized to provide specific applications such as database information access or sales contact management.

Order Processing Systems

Order processing systems gather information related to orders, process the information into specific orders, and create actionable information that allows the fulfillment of the orders. SIP telephone systems can be integrated with order processing systems to allow interactive control with customers to allow the capturing of order information directly from customers and to assist in fulfillment of the order.

Order processing systems within companies are typically limited to data entry from user terminals or by computers connected to the Internet. As a result, order processing systems may require a customer service representative to talk to the customer and enter the order information. This limits the order processing capabilities to the availability of a customer service representative and the potential errors that may result from poor communication skills of the customer service representative and the customer. The use of multimedia SIP telephones allows the user to initiate and enter new orders into an order processing system without the need of a customer service representative.

To enable order processing systems to operate from telephone systems, additional servers or new software on existing servers are added that convert the information from the telephone user (such as keypad entries or audio commands) into commands that can be understood by the order processing system.

Figure 5.1 shows how a SIP based hotel telephone system can be integrated with a hotel's room service order processing system. This example shows how the hotel system has installed SIP-based telephones in each room of the hotel and that each SIP telephone has a display screen. The SIP server and hotel information system (the hotel's order processing system) is connected through the same local area network (LAN) of the hotel. A SIP server is setup to allow users to select and create orders from the room service menu from their SIP based telephone. The SIP server can reformat and deliver the menu order direct to the order processing system. In this example, the order is displayed in the kitchen and the SIP system is used to alert the room service waiter when the order is ready for delivery to the room.

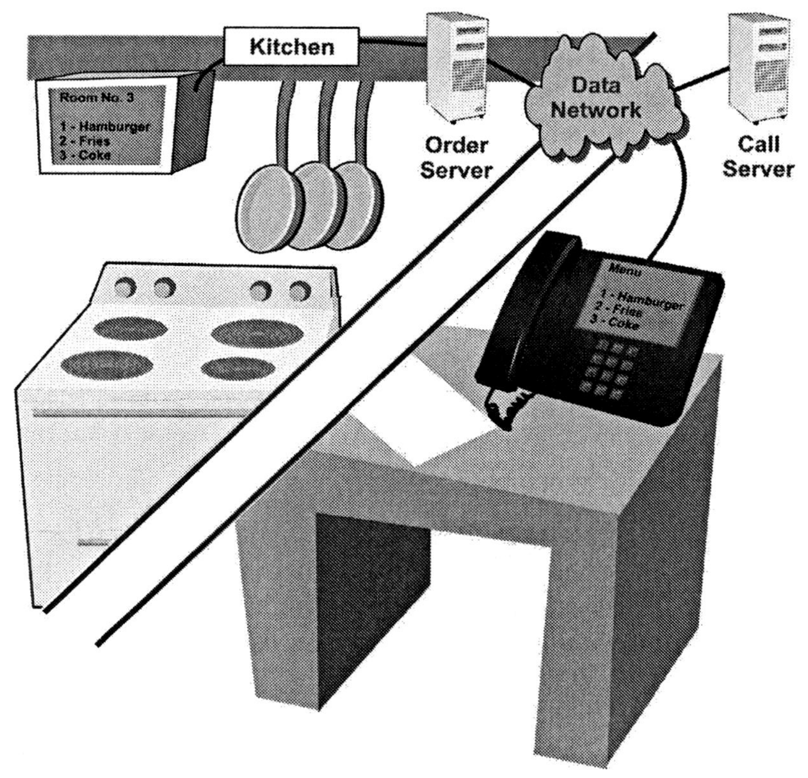

Figure 5.1., SIP Order Processing Operation

Web Servers

Web servers are computer systems that are used provide access to data that is stored and retrieved by commands in Hypertext Transfer Protocol (HTTP). HTTP is a protocol that is used to request and coordinate the transfer of documents between a web server and a web client (user of information). The typical use of web servers is to allow web browsers (graphical interfaces for users) to request and process information through the Internet.

Web servers are limited to providing information to the user in a form and sequence that has been predetermined. As a result, users sometimes need to contact a customer service representative to provide information in an interactive form. Unfortunately, the customer service representative is traditionally limited to audio form. This means the customer has traditionally been limited to using the web page or the telephone to gather the information necessary to purchase a product. SIP systems can be combined with a web server to allow the customer view web pages and communicate through the telephone at the same time.

To combine a web server with a SIP telephone system, the customer should have a multimedia computer with software that is capable of SIP communication and the web server must be modified to establish a communication session with the customer.

Figure 5.2 shows how a SIP telephone system can be integrated with a web server. In this example, a user (potential customer) that has multimedia (audio) capability is accessing a company web page. The user has identified a product the company sells that may satisfy their needs however the user has not found some of the information on the companies web site. In this example, the user selects a "Click to Talk" button and they are connected to a customer service representative. This initiates (invites) a communication session between the user (potential customer) and a customer service representative for the company. The customer service representative can then answer additional questions the customer may have regarding the product purchase. Because a communication session has already been established, the customer service representative can also push information (additional web pages) to the customer that may show the customer that the product performs the necessary features to satisfy their needs.

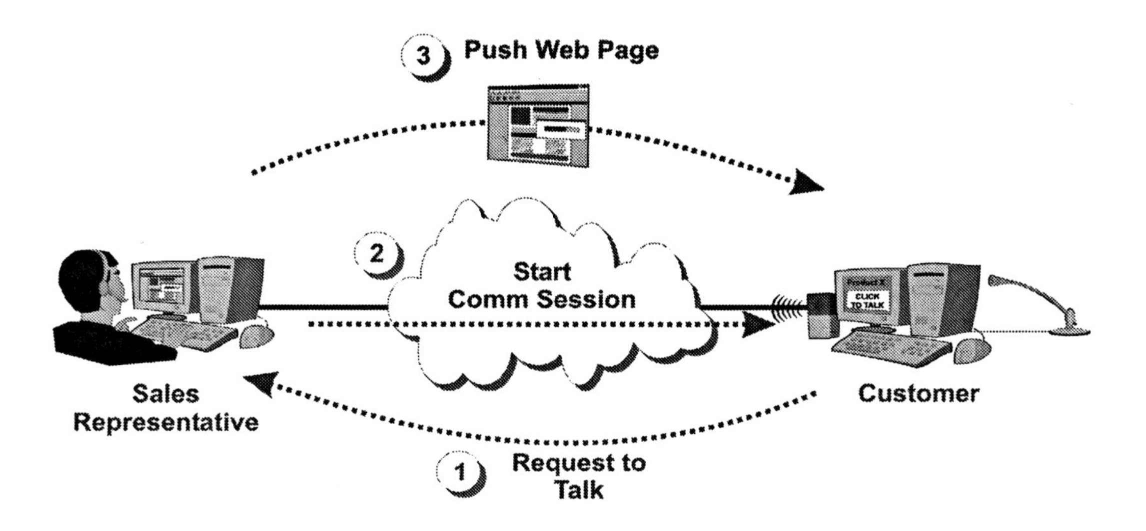

Figure 5.2., SIP Web Server Integration

Instant Messaging (IM)

Instant messaging (IM) is a process that provides for direct messaging connections between computers that are connected to a data communications network. Instant messaging (IM) service usually includes client software that is located on the communicating computers and an instant messaging server that tracks and maintains a list of alias names and their communication status. The IM server usually registers each client and links an address (usually an internet protocol address) so the clients can directly communicate with each other. The client software controls the presentation of information as it is sent directly between each computer.

Instant messaging systems have been traditionally limited to text messages. Because instant messaging systems obtain and share the active IP addresses assigned to the users, it makes it possible to setup voice communication between two or more instant messaging users. Many instant messaging users are familiar with instant messaging software and they have multimedia capable computers so it is a relatively simple process to introduce them to the ability to initiate a voice session.

To upgrade an instant messaging system to use SIP protocol to permit voice communication direct between users, the user's software is upgraded to include SIP protocols. To obtain additional SIP services (such as company directory listings), the user's software must be setup to communicate with the SIP server.

Figure 5.3 shows how an instant messaging system can be integrated with SIP based communication to provide voice service. This diagram shows two people that are instant messaging each other have SIP based voice communication capability. This diagram shows that the IM system has already allowed the participants to discover the IP addresses of the other users. In this example, John uses Barbara's instant messaging address (IP address) to send an invite message to Barbara that requests her to participate in a voice conversation. Barbara acknowledges the request and the instant messaging software negotiates the voice parameters (speech compression in this example) and voice communication session is established.

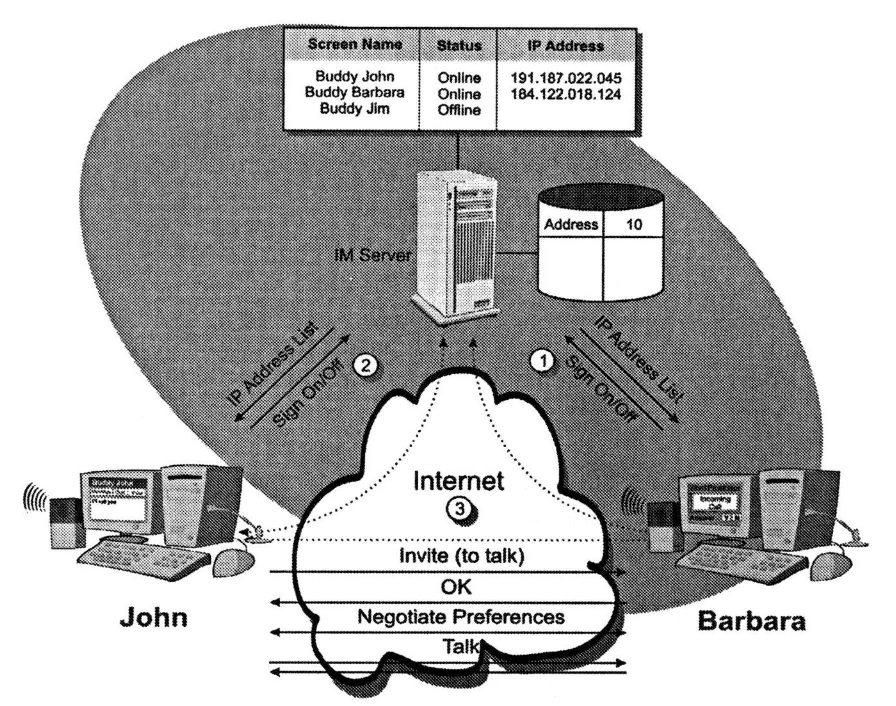

Figure 5.3., SIP Instant Messaging Operation

Web Seminars (Webinar)

Webinars are a seminar or instruction session that uses the Web as a real time presentation format along with audio channels (via web or telephone) that allow participants to listen and possibly interact with the session. Webinars allow people to participate in information or training sessions from any location that has Internet access.

Until recently, the provision of interactive information (such as a training session) to multiple people in different locations required expensive video conferencing facilities or it required participants to travel to common location (such as a conference facility). The use of web seminars allows for the simultaneous provision of audio, video, and data along with the controls necessary for participants to interact with the information moderator.

To upgrade a SIP telephone system to have web seminar capability, a conference server is added. The SIP conference server allows participants and a moderator (typically an instructor) to establish communication sessions with the conference server. The conference server can either receive and forward media from any of the participants or it may allow the participants to directly send media to each other.

Figure 5.4 shows how SIP systems can be used to provide web seminar (webinar) service. This example shows how a presenter can invite several people to participate in a training session. Once the communication sessions (logical paths) are established, the instructor can create additional channels of communication for other multimedia services. This example shows that one of the communications channels is used for audio from the instructor to the students. Another communication channel is used for sending presentation graphics, and a final communication channel is used for sending data files (workbooks).

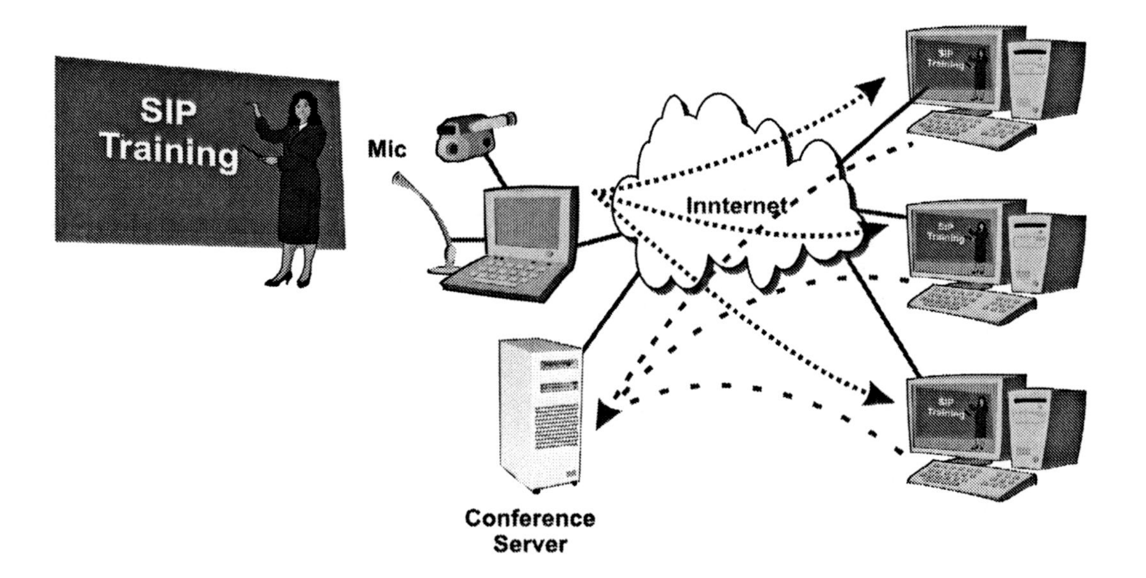

Figure 5.4., SIP Web Seminar Operation

Mobile Communication Information Service

Mobile telephone service (MTS) is a type of service where mobile radio telephones connect users to the Public T Network (PSTN) or to other mobile telephones. Mobile telephone service includes cellular, PCS, specialized and enhanced mobile radio, air-to-ground, marine, and railroad telephone services.

Users desire to have access to multimedia services similar to their computers without the need to connect to wires. Until recently, mobile communication systems have had expensive and limited data transmission capability and the data connection methods have been proprietary. This has forced the system operator (the carrier) to purchase expensive system upgrades and it has not been easy to customize information services for the customers. The use of SIP combined with low cost high-speed mobile telephone digital transmission technology allows customers to have access to new multimedia information services.

To add multimedia services to a mobile communication system, access to a media server is provided by the packet data system. This media server maybe controlled by the service provider (the carrier) or it maybe managed by an independent vendor (information service provider).

Figure 5.5 shows how a mobile phone network can use SIP to add multimedia information services to wireless voice communications. This example shows how a mobile phone with a graphics display can communicate on the high-speed packet data communication channel to a SIP server to obtain driving direction information. In this example, a mobile telephone user has requested a session with a company that provides driving direction information services. When this user requests a connection (sends an invite), the user is first validated as a subscriber of the map information provider. After the customer's account has been validated, a communication service (logical connection) is established between the mobile device and the media server. The user will browse through a menu that allows them to set the parameters (starting and destination address) and the media server can create the information and graphics that are transferred to the mobile device.

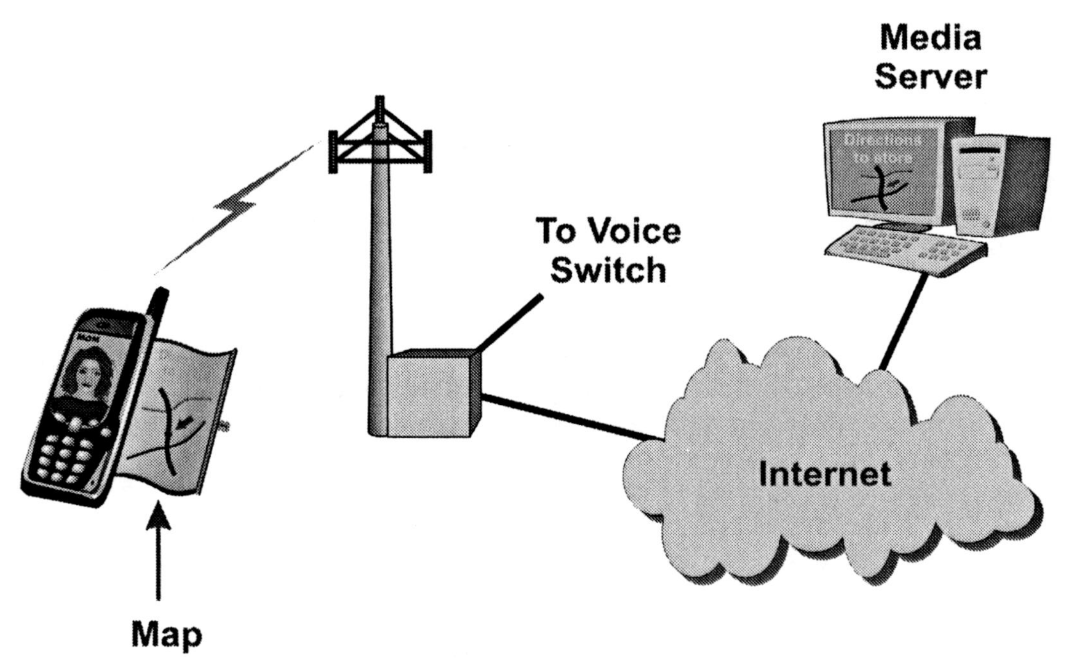

Figure 5.5., SIP Mobile Communication Operation

Database System

Databases are collections of data that is interrelated and stored in memory (disk, computer, or other data storage medium.) Database systems are typically accessed and controlled by computer terminals that are connected to the same data network as the database system.

The information contained in corporate database systems is typically only accessible through computer terminals. However, database information is commonly stored in a standard form such as Structured Query Language (SQL) format. This allows other servers to access, sort, and retrieve information by using commands that use the standard data storage format. To integrate a database system (such as a customer database) with a SIP based telephone system, various types of servers (such as a voice recognition server) can be added, these may allow for telephone users to control the access to the database by voice commands.

Figure 5.6 shows how a SIP based phone system can be integrated with a company's customer database system. This example shows how a sales person can use a standard telephone to call into a company SIP based IP PBX system and link the call to the company's information order processing database system. This example shows that the company sales representative dials into the IP PBX system and connects to a voice recognition server (possibly an extension with an access code). This allows the salesperson to simply say what he is requesting and this information is translated to the commands that are filtered through an application server that can in turn communicate with the company's existing (legacy) order processing system. When the sales person requests "order status" "Customer" "Best Customer," the application server (AS) creates the necessary commands to link to the company's order processing database, search for a specific customer, and run a query on what orders are being processed. The sales person can then navigate through the order processing systems by issuing verbal commands that are translated through the SIP system to the order processing system.

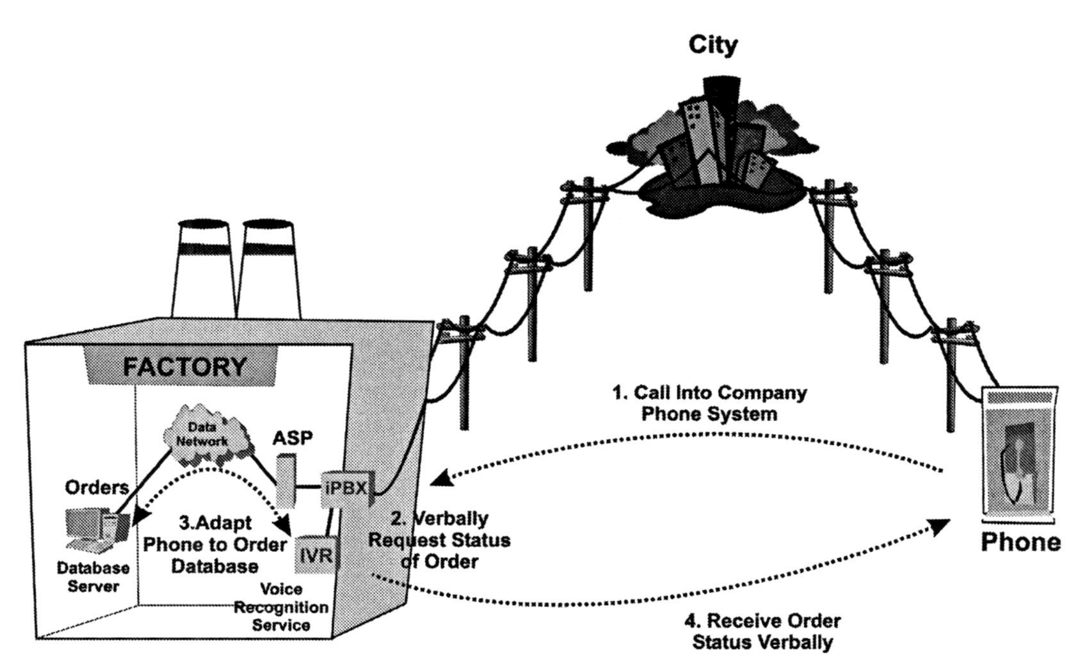

Figure 5.6., SIP Database Integration Operation

Dispatch Systems

Dispatch systems provide radio service that allows a central controller (dispatcher) to send dispatch assignments to one or more receivers (typically many mobile radios). Dispatch radio systems normally involve the coordination of a fleet of users via a dispatcher. All mobile units and the dispatcher can usually hear all the conversations between users in a dispatch group by setting the users to a channel (or channel code) that is shared by all the users in the group. Dispatch operation involves push-to-talk operation by a group of users on the dispatch system. Using radio trunking (multichannel access) technology, there can be several different dispatch groups that operate (share) on the same system.

There are many dispatch radio systems in use that are incompatible with each other. Some of these dispatch systems use analog radio, some use proprietary digital formats, and some use digital radios that can transmit standard Internet protocol (IP) packets. By converting or adapting the radio channels of dispatch radio systems to a standard IP telephony format that uses standard SIP protocols, it is possible to interconnect incompatible dispatch radio systems over wide geographic areas.

Figure 5.7 shows how multiple dispatch radio networks can be integrated using a SIP system. This example shows that different types of dispatch networks have been interconnected by high-speed packet data networks. The analog land mobile radio dispatch systems are connected through a gateway that converts the analog voice and control signals to digital messages. Digital land mobile radios that use a proprietary communication system are connected through a gateway that converts the digital audio from one format to a format that is compatible with the IP dispatch system. Digital land mobile radios that use IP packets and SIP protocol can be directly connected to the data network.

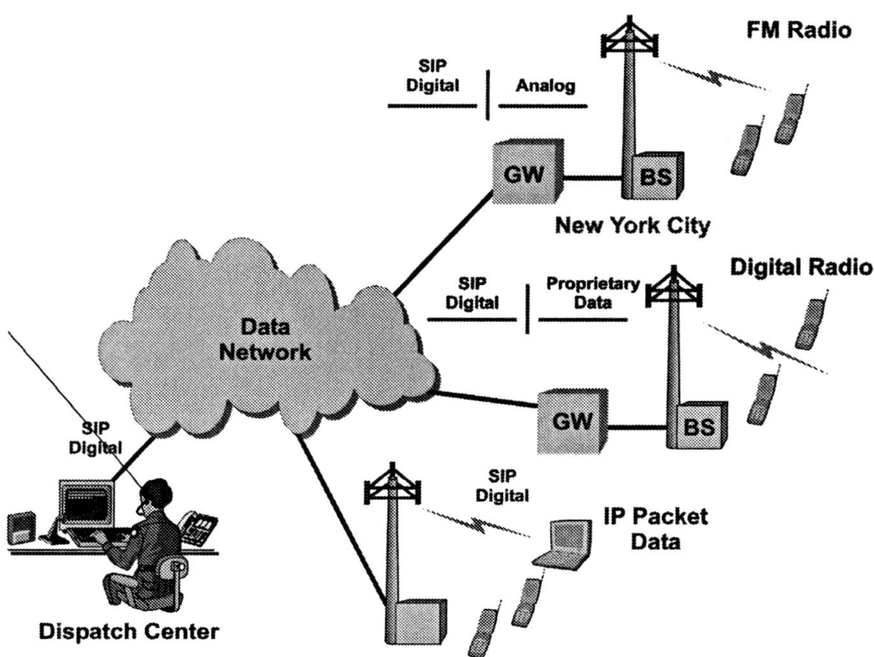

Figure 5.7., SIP Dispatch Operation

Security Systems

Security systems are monitoring and alerting systems that are configured to provide surveillance and information recording for protection from burglary, fire, safety, and other types of losses.

Traditional (legacy) security systems use proprietary sensing and transmission equipment, have limited control processing capabilities, and have interconnections that are limited to local geographic areas. The use of SIP systems connected through standard data networks allows for the sending of media (such as digital video), powerful security system processing in a server, and wide area connectivity (such as through the Internet).

Figure 5.8 shows how a variety of security accessories can be integrated into a SIP based communication system. This example shows how a police station can monitor multiple locations (several banks) through the addition of digital video and alarm connections. This example shows that when a trigger alarm occurs at a bank (such as when a bank teller presses a silent alarm button), the police can immediately see what is occurring at the bank in real-time. Because the images are already in digital format, it may be possible to send these pictures to police cars in the local area to help identify the bank robbers.

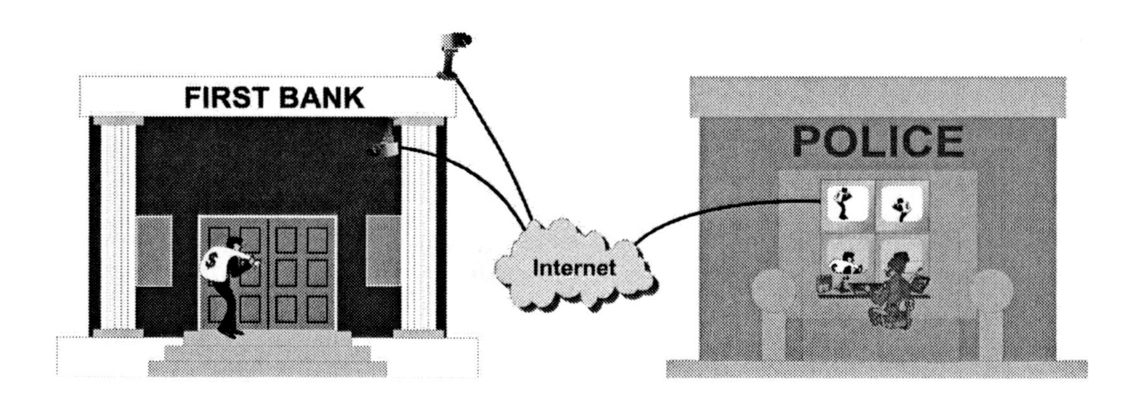

Figure 5.8., SIP Security System Integration

Interactive Television

Interactive television (iTV) is the combination of video service with the ability to dynamically alter the content or flow of a media program that is provided to a user. Until recently, television and cable television have been delivered in a one-way broadcast process. One-way analog cable television video systems have been transitioning to two-way high-speed digital communication systems.

Cable television systems are composed of a head-end system (the network television receivers), the distribution network, and end user equipment (set top boxes). Until recently, the set top boxes have been proprietary and the control of the system has been limited to sequential programming and non-changeable media. The use of SIP based systems and a digital media server allows the user to change the programming dynamically based on user inputs and/or predetermined preferences. This can provide preferences such as the elimination or skipping of sections of bad language, nudity, violence or even the selection of different outcomes to a movie (positive or negative). To add SIP capability to digital cable television systems, the users need to have terminals with SIP capability (Multimedia MPEG to NTSC or PAL) and a media server at the head end of the cable television network.

Figure 5.9 shows how a cable television network can use SIP to add interactive media services to video communication networks. This example shows a video on demand (VOD) system that uses SIP to initiate and manage video delivery from a digital movie storage system. In this example, the cable television network has two-way digital communication service. The user is provided with a digital set-top box and the cable television network has installed a media server and an application server that can receive and process requests from users. This example shows that the user has a remote control that can send commands (SIP messages) to the media server at the head-end of the cable television network. These commands allow the end user to control the media source (such as play, stop, or rewind).

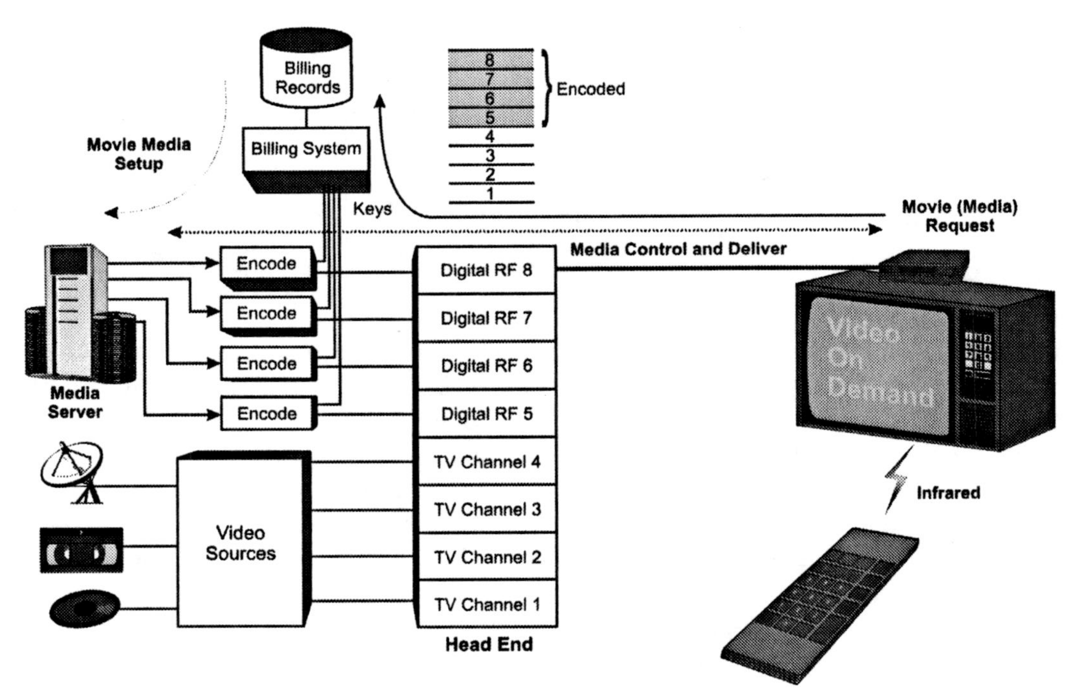

Figure 5.9., SIP Cable Television Operation

Chapter 6

Building SIP Systems

The set up of a SIP system involves the design of a communication network and purchasing SIP equipment and core software. It may also involve the purchase of SIP applications software (such as voice mail modules) and possibly the development of software for custom system features.

Some of the reasons for installing a new SIP-based communication system include replacing equipment that is near its end of useful, the extension of telephone systems to remote sites, the installation of equipment at new locations, or the integration of telephone systems with information systems.

Designing a SIP system can range from a simple turn-key solution to a complicated system integration process. A key option for the setup of a SIP system is to use (or start with) a hosted SIP communications system (called IP Centrex) instead of purchasing and operating a complete SIP system.

Building a SIP system involves the setup of one or many computer servers. Servers are referred to by different names by different manufacturers. These servers include call managers (proxy servers), system administrators, unit managers (location servers), and gateway managers. A system manager may be used to link all of the different servers to each other. After the servers have been setup, dial plans are created and advanced call processing features may be created (such as call groups).

SIP System Design

SIP system design involves identifying the key functional requirements of the desired system, evaluation of the data network requirements, review of the existing telephone system, analysis of available options, and the layout of the new system.

One of the first steps in SIP system design should be the definition of the functional business requirements for the new communication system. These functional requirements usually include traditional PBX voice and advanced call-processing features. They also may include unified messaging, call center, and security system integration features.

The first stage of system design is to make diagrams of your existing data and telephone systems. This should include cable diagrams along with the types of cables (number of lines and category of lines) that are used. This will allow for the reuse of existing cabling when possible.

If an existing data network is going to be used as part of the SIP system, it should be evaluated to determine if it has the available capacity (bandwidth), reliability, and quality of service (QoS) capability desired for the new SIP system. The bandwidth should be evaluated at concentration points (routers and switches) along with the capacity of wide area connections that will be used as part of the SIP communication system. The reliability objectives of the system should be considered and alternative routes and backup power supplies may need to be added. The quality of service (QoS) requirements of data communications and voice systems may be considered and a policy server may be used to give priority to voice packets.

The analysis of available options include which protocols will be deployed (such as SIP, MGCP, or H.323), the use of a partial or complete hosted (IP Centrex) solution, the use of separate networks (independent phone and data), or an integrated network (shared data system).

Once the existing systems have been evaluated and the system options of protocols, server platforms, security firewalls, and end-user equipment types have been selected, layout of the system can begin. The system layout design will usually include the servers (computers) and their names, cabling types, end-user devices, and possibly software programs. The layout may also include connections and references to other carriers (such as an IP Centrex provider).

Figure 6.1 shows the basic design of a SIP system. This diagram shows that the SIP system is composed of servers (computers), end user devices (IP telephones), gateways (media adapters), and interconnecting data networks. This diagram shows a call manager server that coordinates the origination and reception of calls between telephone devices in its area (it's domain). An

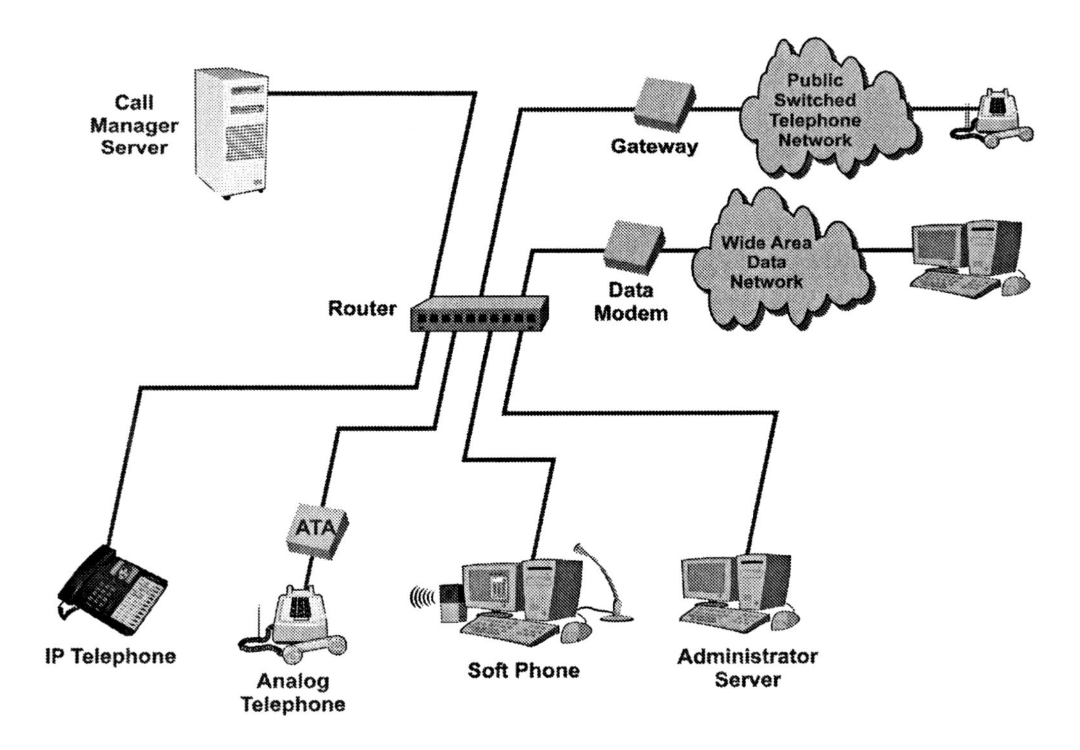

Figure 6.1, Basic SIP System Design

administrative server attached to the network that is used to setup and manage user accounts on the system. This system includes multiple types of phones including IP telephones, analog telephones (through an analog telephone adapter), and softphones (on multimedia computers). The call manager can complete calls through a wide area data network or through a voice gateway that connects this system to the public switched telephone network.

Hosted SIP Systems

Hosted SIP systems are communication systems that divide between end user terminals and the call processing hardware and software that is managed by an external company. The use of a hosted SIP system allows a company to benefit from the flexibility of SIP technology and integration without the need to understand or manage the call processing. There are two basic types of hosted systems; Internet Telephone Service Provider (ITSP) and IP Centrex.

The use of a hosted system simplifies the development of SIP based communication systems as the IP Centrex provider performs much of the call control and equipment management functions while allowing the customer to add, modify, and delete users from the system. Hosted systems typically provide an administrator access point (a web portal) to allow account management for the system services.

Internet Telephone Service Provider (ITSP)

Internet Telephony Service Providers (ITSPs) are companies that provide telephone service using the Internet. ITSPs setup and manage calls between Internet telephones and other telephone type devices.

An ITSP coordinates Internet telephone devices so they can use the Internet as a connection path between other telephones. ITSPs are commonly used to connect Internet telephones or PC telephones to telephones that are connected to the public telephone network at remote locations.

Figure 6.2 shows how an ITSP sets up connections between Internet telephones and telephone gateways. The ITSP usually receives registration messages from an Internet telephone when it is first connected to the Internet. This registration message indicates the current Internet Address (IP address) of the Internet telephone. When the Internet telephone makes a call, it sends a message to the ITSP that includes the destination telephone number it wants to connect to. The ITSP reviews the destination telephone number with a list of authorized gateways. This list identifies to the ITSP one or more gateways that are located near the destination number and that can deliver the call. The ITSP sends a setup message to the gateway that includes the destination telephone number, the parameters of the call (bandwidth and type of speech compression), along with the current

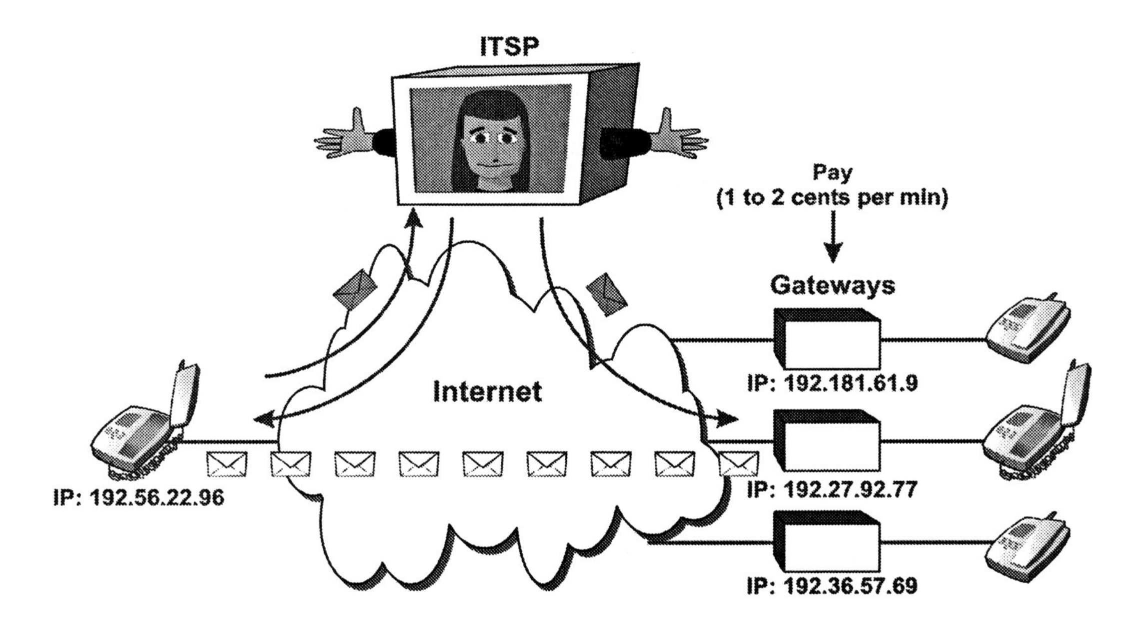

Figure 6.2, Internet Telephone Service Provider (ITSP)

Internet address of the calling Internet telephone. The ITSP then sends the address of the destination gateway to the calling Internet telephone. The Internet telephone then can send packets directly to the gateway and the gateway initiates a local call to the destination telephone. If the destination telephone answers, two audio paths between the gateway and the Internet telephone are created. One for each direction and the call operates as a telephone call.

IP Centrex Operators

Centrex, (a contraction of the term Centralized Exchange), is a telecommunications service that allows customers to get the full range of features available on a PBX, without actually owning or renting PBX hardware. The Centrex services are delivered to the customer by a dedicated partition of the central office or local exchange that behaves like the customer's PBX. IP Centrex operators provide Centrex like services to customers using Internet Protocol (IP) connections.

Figure 6.3 shows a basic IP Centrex system that allows a Local Exchange Company (LEC) in New York City to provide Centrex services to a company in Los Angeles. In this diagram, the LEC in New York City uses a Class 5 Switch to provide for Plain Old Telephone Services (POTS) and Centrex services to their local customers. The Centrex software is installed in the switch and existing Centrex customers in the local area continue to connect their telephone stations directly to the Class 5 switch. To provide Centrex services to new customers located outside the geographic area, the LEC has installed a network gateway in New York that can communicate with the customer gateway in Los Angeles. Because the network gateway converts all the necessary signaling commands to control and communicate with the customer gateway, the Class 5 Switch does not care if the customer gateway is in Los Angeles or Tokyo. It simply provides the Centrex services as the user's request.

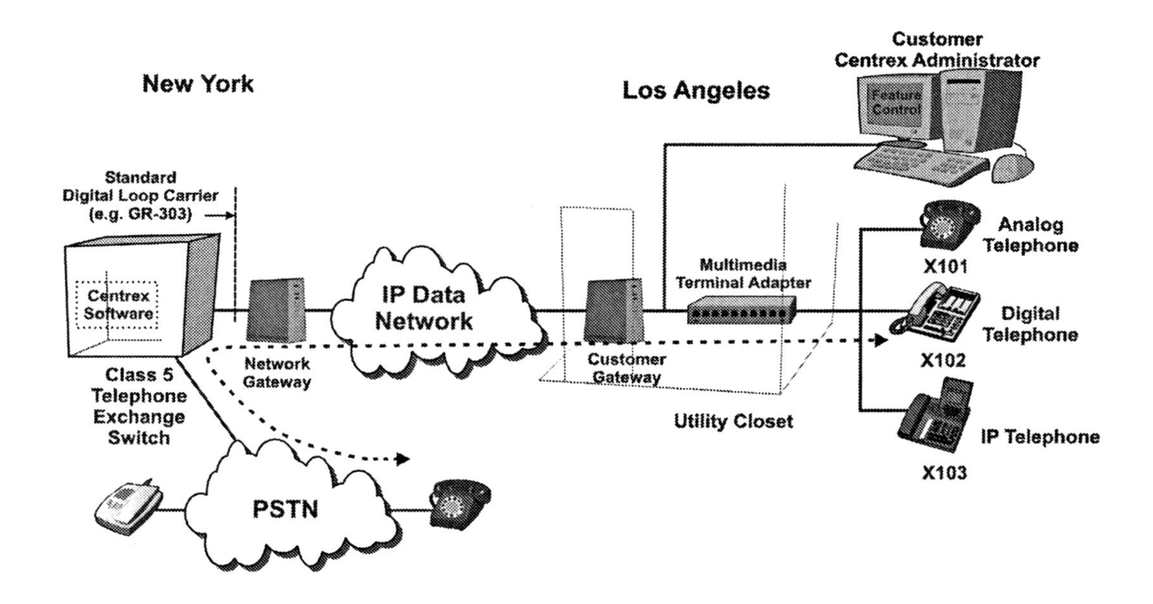

Figure 6.3, SIP Hosted (IP Centrex) Communication System

SIP Communication Servers

SIP communication servers are the core of the SIP communication system. SIP servers within a network can either be located on different machines or can be located on a single machine. When the communication servers are co-located on the same computer, they use different port addresses (separate logical channels). Servers are computers with a server operating system. Popular server software systems include Windows NT, Windows 2000 Server, or Linux. While it is possible to setup a SIP-based communication system on a computer that is used for other functions, it is common to have a dedicated computer for the SIP communication server.

SIP communication servers receive request for calls, assist in the setup of calls, perform call feature operations during the call, and process requests in the termination (ending) of calls. SIP communication servers may translate standard telephone numbers into their associated IP addresses, look for available gateways to complete calls between the PSTN and SIP devices, and manage advanced communication features.

Manufacturers and vendors commonly use server names that may be different than the names assigned by the industry standards committees. For example, a marshal server or a call manager is a proxy server and a unit manager may function as a location server.

Figure 6.4 shows the different types of servers used in some SIP based communication systems. This example shows that a call manager (proxy server) receives and processes call requests from communication units (IP telephones). The administrator server coordinates accounts to the system. A unit manager (location server) functions as a location server by tracking the IP address assigned to the communication units. The gateway manager identifies and coordinates communication through the available gateways. The system manager coordinates the communication between the different servers and programs available on the system.

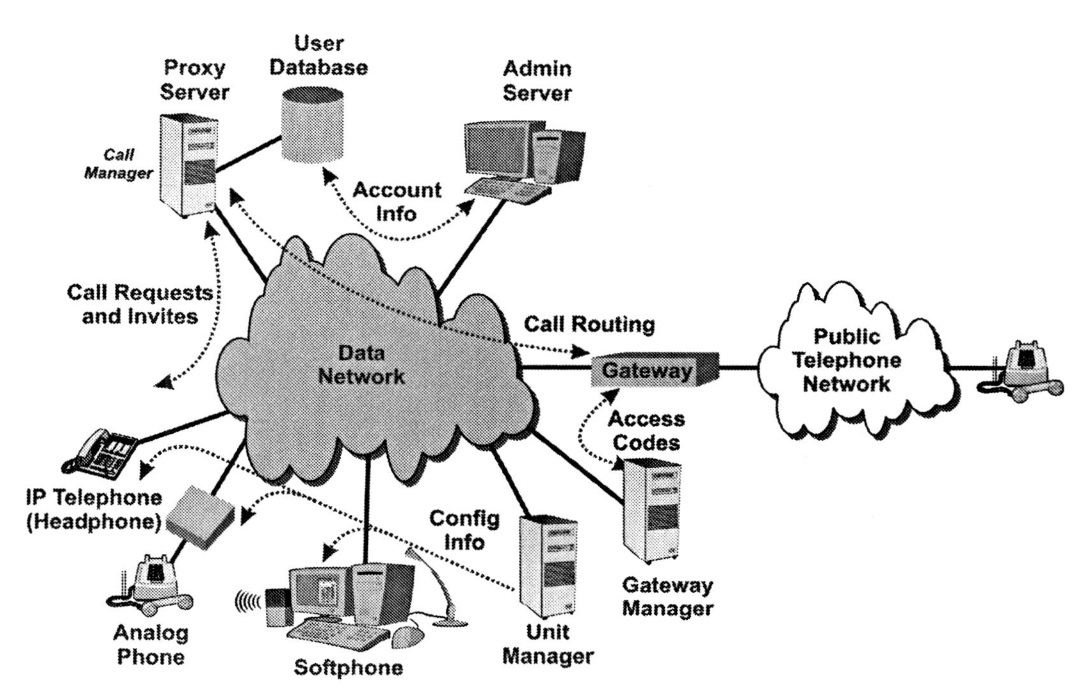

Figure 6.4, SIP Communication Servers

SIP Administrator

SIP administrator software is graphic user interface (GUI) for the system administrators to setup and manage accounts on the SIP system. The administrator has the ability to move, add, and change (MAC) accounts. The setup of the system administrator is one of the first steps in building a SIP system.

Because SIP communication systems are interconnected by IP data networks that are typically connected through the Internet this allows administrators to setup and manage their system from any place they can connect to the Internet.

Figure 6.5 shows the basic functions of the SIP system administrator. This diagram shows that the SIP system administrator is responsible for adding new accounts to the SIP system, assigning the dialing rights (such as international and long-distance dialing), and the features authorized (such as conference calling). This diagram also shows that the SIP system administrator can perform these assignments and changes from a web portal.

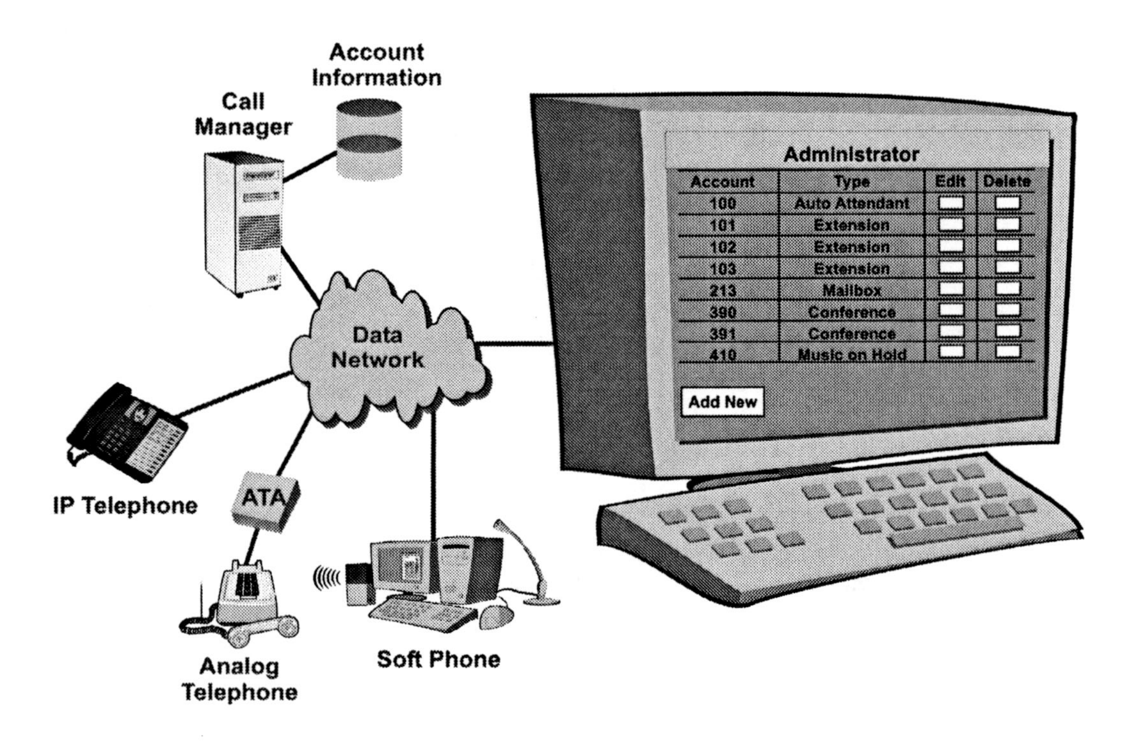

Figure 6.5, SIP Administrator Software

Call Manager

A call manger (SIP proxy server) is the call control processing software that can receive call requests (call invites) from users and assist in, (proxy on behalf of), the setup of connections between communication devices. Call managers only help setup the call connections, they do not actually transfer the call data.

The call manager server may translate SIP commands to other protocols such as Media Gateway Control Protocol (MGCP) or H.323. This allows for a SIP based system to use devices (such as Voice Gateways) that do not use SIP protocol. In some cases, a separate translation server may be used to convert protocols to and from devices that use other protocols.

Figure 6.6 shows common setup of the call manager (call proxy server) software for a SIP communication server. This diagram shows that the SIP server port (logical) channel is commonly set to 5060 (the default). This diagram also shows that the SIP call manager can be setup to use an alternative another server that coordinates the end-user devices. This identifies the IP address of the user manager server. The example shows that authentication security can be setup on another server.

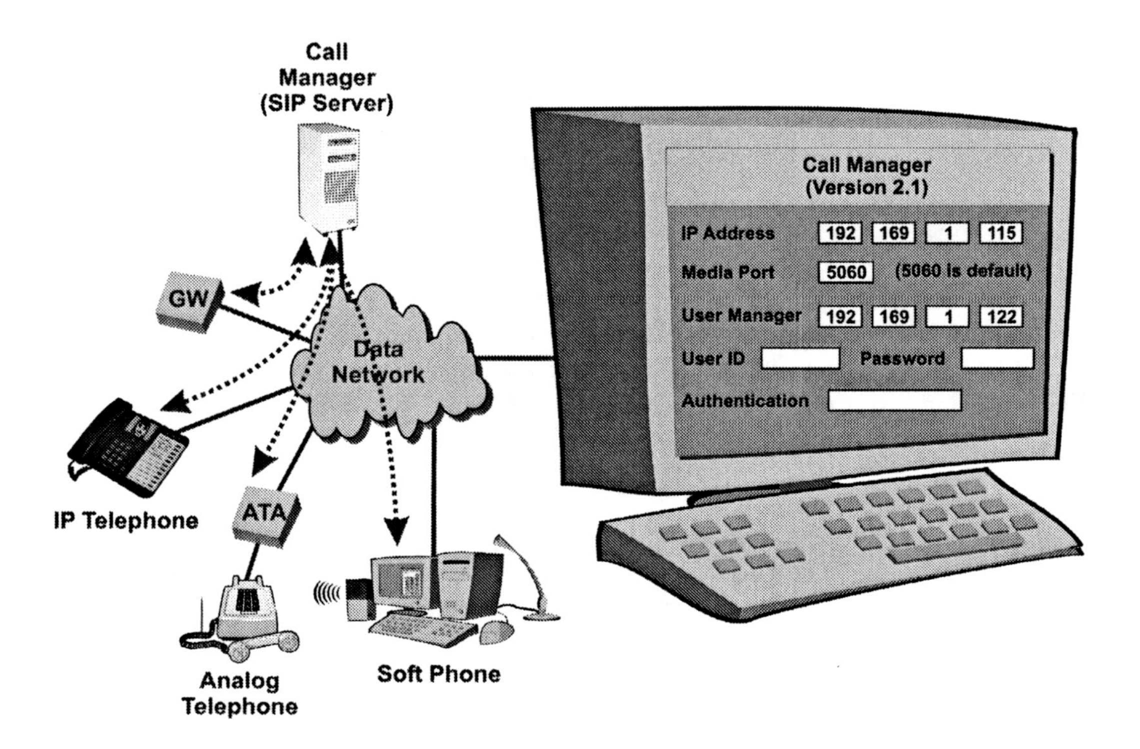

Figure 6.6, SIP Call Manager Server (Proxy Server) Software

Gateway Manager

A gateway manager is used to configure gateways that connect data networks to other networks such as the public switched telephone network (PSTN). The gateway manager keeps track of the available gateways and what devices in the system are allowed to connect to them. Gateways maybe owned by the user or they may be available through other companies on a fee basis

The gateway manager will setup the authentication and access rights for each gateway. To allow services through the gateway, the IP address of the calling device or system may be used along with account codes and passwords.

Figure 6.7 shows how gateway manager software can be used to configure and manage gateways that connect the SIP network to other networks such as the public switched telephone network. This example shows how the gateway manager contains the configuration information for the gateway including IP address, capabilities such as speech coders, protocols, and access control information.

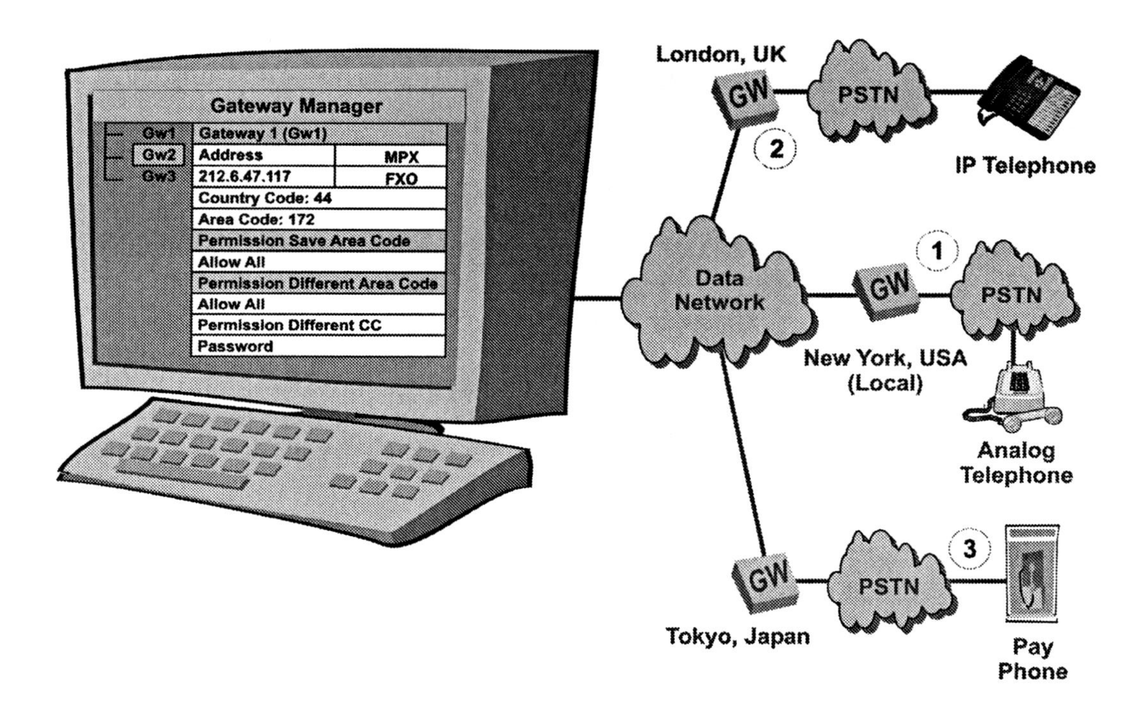

Figure 6.7, SIP Gateway Manager Software

Unit Manager

A unit manager server is used to setup (configure) and map (identify) the communication units (SIP telephones and media gateways) to the system. IP Telephony units require configuration so they can initially find the server (their addresses) that will coordinate with and to know the protocols (such as SIP) and processes to use (such as the type of speech compression used).

Units within the network may be manually configured through the unit manager or they may be configured automatically through automatic detection (device discovery) and downloading of configuration files. Unit configuration may also be done without the use of a unit manager server by logging onto the units IP address using a web browser (passwords are usually required) or by editing parameters of the Management Information Base (MIB) configuration file of the Unit using Simple Network Management Protocol (SNMP) commands.

Units are uniquely identified by an IP address (network address) and a MAC address (physical address). The IP address assigned to a unit may be fixed (static) or it may be assigned after it is turned on (dynamic addressing). MAC addresses are physically programmed into the device at the time of manufacture.

Units will usually be setup to communicate with one server or a group of servers to allow it to use different types of services (such as internal calls, incoming calls, outgoing calls, and conference calls). The unit may be programmed to search for its servers (to register) when it powers on. Unit configuration includes setting the addresses of servers used by the device (proxy servers), the identification and authentication information, the dialing plan and preferred features, when server registration will occur, and how new operating software and configuration data may be transferred to the device.

The SIP server address is the proxy server that will handle the registrations and other call setup functions for the device. The outbound proxy setting is the server address that will handle call origination requests. The outbound proxy may convert the dialed digits of the telephone number into an IP address of the gateway that will connect the call from the data network to the public telephone network. The SIP user ID is the SIP identification (possibly the account ID) that this device will use during its call setup requests. The Authenticate ID is the identification code used by the device when it is required to authenticate itself to obtain service. The authenticate ID is sometimes the same as the SIP user ID. The Authenticate password is used as part of the information processed the authentication process to determine if the device is authorized to request and receive services. A name (such as a user name) may be programmed into the device to allow the recipient of a call request to see who is calling (caller identification).

Part of the setup usually includes the preferred features such as the preferred speech compression type and how often SIP registration occurs.

The unit's operating software (firmware) may be updated through the use of transferred files by means of the Trivial File Transfer Protocol (TFTP). TFTP is a slim version of a file manager system. The downloading of new firmware may be necessary to fix incorrect operations (software bugs) or to provide for new feature capabilities to the units.

Figure 6.8 shows a typical unit configuration screen that is used to manually program the settings into an end-user device. This screen shows that the configuration of a SIP unit involves setting the IP address mode (dynamic DHCP or static), setting a default router address, entering the address of the DNS server (for text address to IP address conversion), and the addresses of the call manager (SIP proxy server). The setup also includes identifying information such as the SIP user ID, login ID, and passwords. Optionally, some preference options may be set such as speech coder type (high or low bandwidth), silence suppression, registration periods, and dial strings.

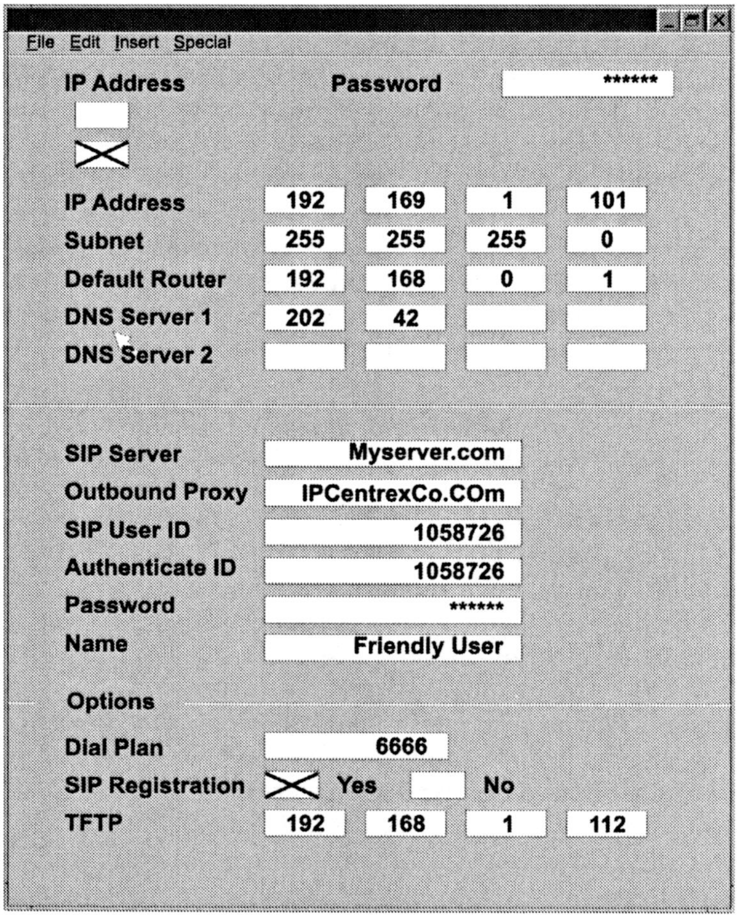

Figure 6.8, SIP Unit Configuration Screen

When using a unit manager server to control the setup of SIP devices, the devices that are managed usually include their status in the network (configured, off-line, active). The status is usually indicated by different colors. Red usually indicates that the unit is not configured or is inoperable. Blue may indicate that it has received some programming. Yellow might indicate the unit is partially operational. Grey can indicate the unit is ready (correctly configured) to operate.

Figure 6.9 shows how SIP system unit manager software can be used to setup and maintain the configuration settings for end-user and gateway devices that can be managed by the SIP system. In addition to maintaining the configuration, the Unit Manager will usually indicate the status of each unit that is connected to the network. This diagram shows that this unit manager provides different symbols for each type of device (such as an analog telephone adapter or an IP telephone) that are connected to the system. This system also changes the color of each device icon depending on its status (new, configured, online, or in-use).

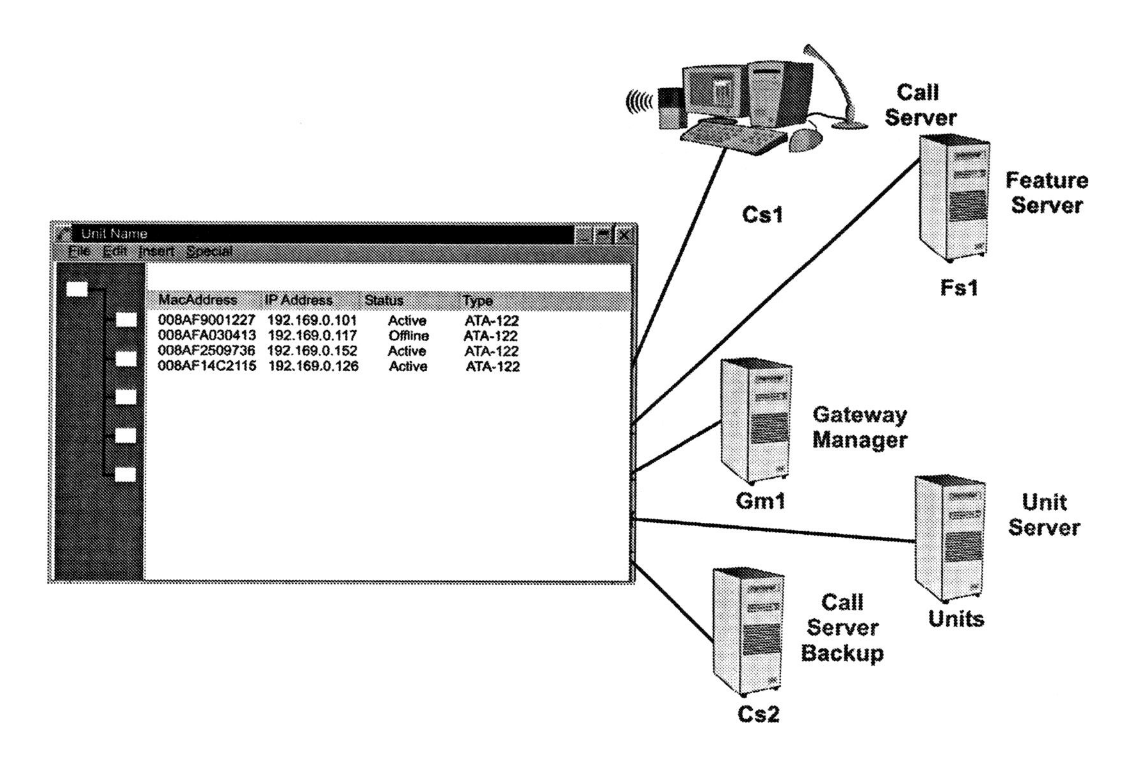

Figure 6.9, SIP Unit Manager Software

To uniquely identify each unit device, the device has an IP address (that may dynamically change) and a MAC addresses (that is pre-set at the factory and does not change). The type of device may vary from complete IP telephones (hardphones) to software telephones on multimedia computers (softphones).

To identify each device and its capability in the network, all the devices in the network may be requested to register. The addresses of the SIP devices can be manually entered into the unit manager or they can be automatically detected. To automatically detect if a SIP device is connected to the system, the Unit manager can send registration requests to all the devices in its network. To keep the amount of data transmission low and to keep from confusing non-SIP units, registration requests can be sent within a range of addresses. This means that the addresses of end-user devices should only be programmed within a specific range of addresses. For example, 192.169.0.100 to 192.169.0.255.

Figure 6.10 shows how SIP unit device automatic detection software can be used to automatically find end-user and media gateway devices. This example shows that the administrator has entered an IP address search range of 192.169.0.100 to 192.169.0.255 and that the software has found 4 devices. The search result produces the IP address, MAC address, and the type of device that has been found.

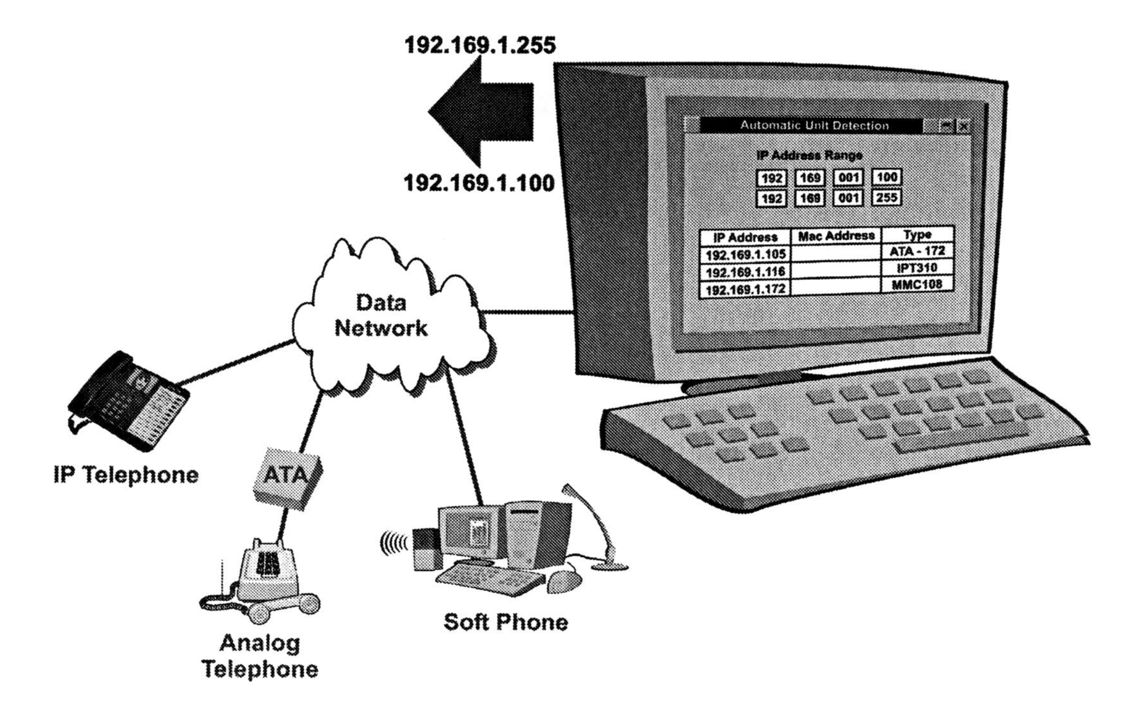

Figure 6.10, SIP Unit Autodetection Software

System Manager

A SIP system manager server controls and links all of the SIP system elements (servers) to each other. The system manager coordinates the routing of SIP messages through the system by assigning addresses and communication ports (logical channels) to each server.

It is common to assign a short mnemonic name to each communication server. This allows for quick identification of the server and the type of service it primarily performs.

Figure 6.11 shows how system manager software is used to link all of the system server components to each other. This example shows a communication server (named CS1) that is operating as a proxy server. A feature server (called FS1) is setup to perform advanced feature processing. The gateway manager (named GM1) is used to coordinate calls between the SIP data network and the public telephone network. The unit server (called Units) is used to manage the communication with specific types of end-user equipment. An additional server called CS2 is used as a backup communication server for the CS1 proxy server.

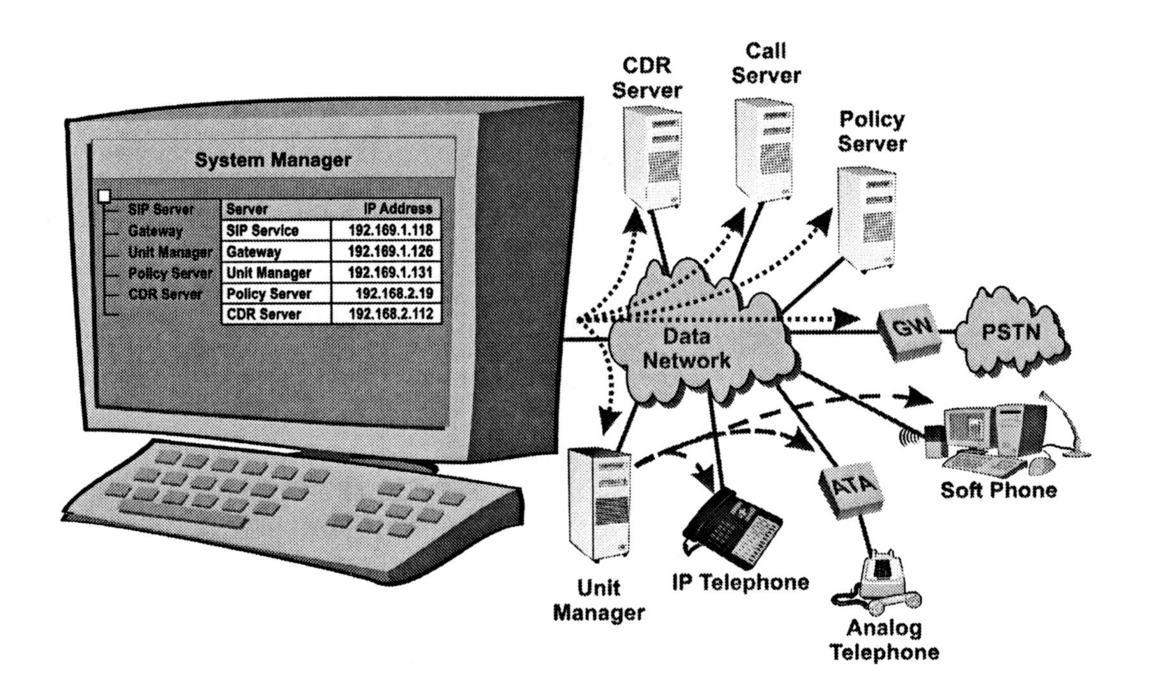

Figure 6.11, SIP System Manager Software

Translation Server

A translation server in a SIP system is a signaling gateway that can convert from one protocol to another protocol. While the IP Telephony protocols are different (for example SIP, H.323 and MGCP), they perform similar functions. Some of the commands used during the setup, management, and termination of calls require the knowledge of the call status (the state). This means a translation server performs more functions than translating one message to another message.

User Manager

A user manager stores the features that have been selected by end users. These features may include speed dialing, call forwarding, voice mail options, and other configuration settings the user is allowed to select. To access the user manager feature, the end-user typically logs into the user manager web screen using their account and password.

Figure 6.12 shows some of the feature options a user manager may display to a user. This example shows that each user must first log into their account using their account ID and password. The user can then select to setup options such as call forwarding, voice mail, email alerting, and other features.

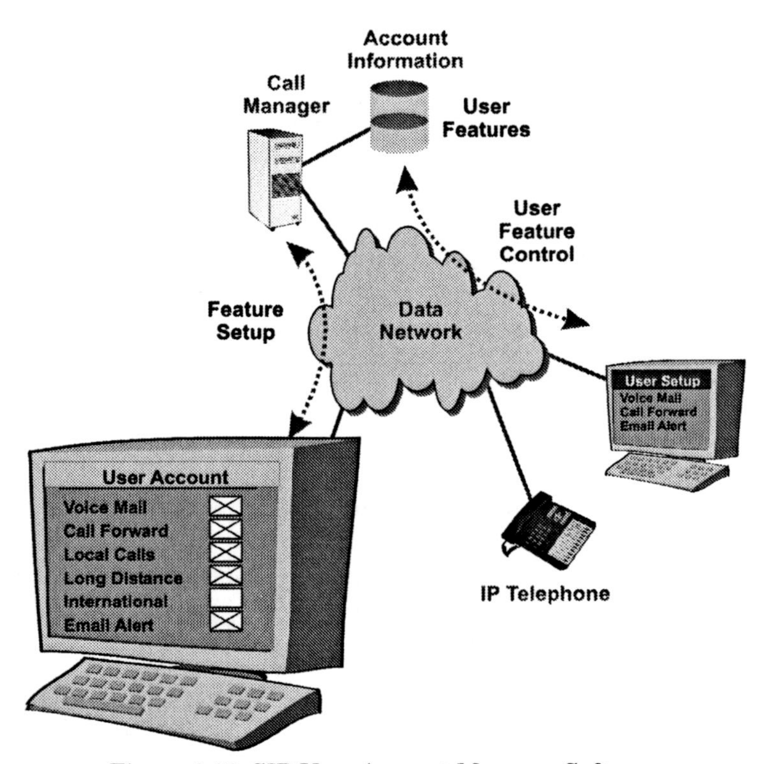

Figure 6.12, SIP User Account Manager Software

Conference Server

A conference server (bridge) is a telecommunications facility or service that permits callers from several diverse locations to be connected together for a conference call. The conference server in a SIP communication system must perform two key functions: controlling the addition and removal of participants and media mixing.

The conference bridge contains electronics for amplifying and balancing the loudness of each speaker in a conference call, so everyone can hear each other and speak to each other. Background noises are suppressed and typically only the current two or three loudest speakers' voices are retransmitted to other participants by the bridge, while a speaker's own voice audio is not sent back to that speaker to avoid audio feedback, echo or "squealing" self-oscillation.

Conference servers can be setup to manage one-step or two-step conference connections. One-step conference connections allow a caller to be connected directly to a conference by a calling a single number or IP address associated with the conference session. The two-step conference connection occurs when the caller first dials a universal phone number or IP address associated with conferencing and then uses an access code to connect to a specific conference.

Figure 6.13 shows how a SIP system can use a conference server can be used to receive requests to allow connections to a conference media server. This example shows a two-step process that has assigned a single telephone number for conference calls. When the conference server receives the connection request, it redirects the call to an interactive voice response (IVR) unit. The IVR prompts the caller to enter the conference identification information. When the IVR has collected the appropriate conference identification information, it redirects the call to the conference bridge.

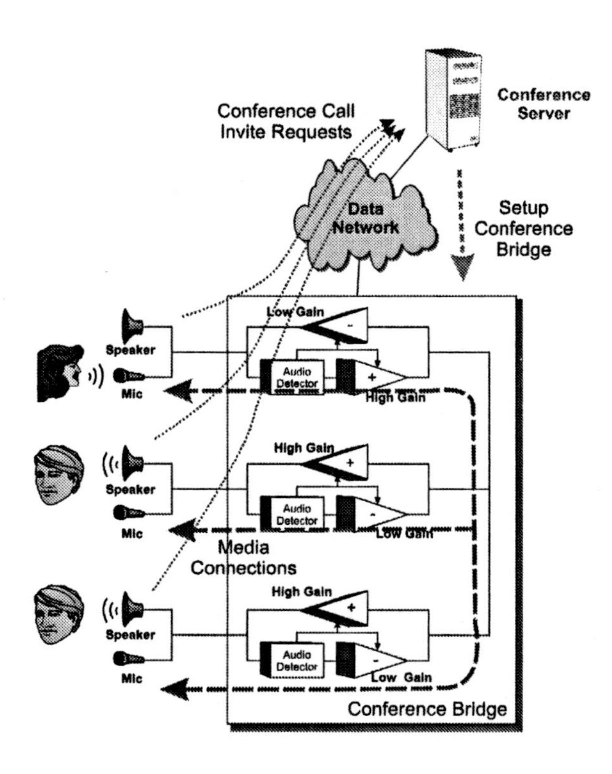

Figure 6.13, SIP Conference Server

Dial Plan

A dial plan (also called a dialing scheme) is the numbering system that is used by a company to identify devices within their network by unique numbers. After a system has been setup, a dialing plan is developed for each communication unit (or groups of communication units).

To implement a dial plan, a dial map is used. The dial map is the systematic use of certain prefix digits to dial a destination via user selected routing. An example is the use of the dialed prefix "9" from within a PBX to first select an outside local telephone line so that the originator can then dial a (typically 7 digit) local city telephone number. Similarly, a PBX may use the dialed prefix “8” to select a tie line to another PBX. A dialing plan differs from a numbering plan by being used inside a particular private telephone system, and also the specifics of different dialing plans are different for different PBXs or private networks in the same country, while a numbering plan is uniform throughout an entire country.

The dialing map must take into account dialing for emergency services request. Different countries use different numbers for emergency services (such as 911 in the USA and 999 in the UK). Each device within a SIP system can be setup to use a different calling map to adjust for different calling patterns.

Figure 6.14 shows how a basic dial map operates. This diagram shows that there are several dial plan rules that are used each time a number is dialed. The first step in the dial map is to determine if the first digit is a 0, 3, 8, or 9 are dialed. This first rule allows the system to determine if the caller desires to reach the attendant (0), is calling an internal number (3+), long distance (8), or outside line or emergency services (9). This rule changes how the next digit is processed. If the first digit is a 3, it is an internal call (4 digits for this system) and the system will wait for 3 more digits before attempting to connect the call to another unit in the system. If the first digit is 8, the system will capture multiple digits and analyze the call as a long distance public telephone number (country code, city code, exchange code, and extension). If the first digit is a 9, the system will analyze the following digits to determine if it is an emergency call (for example 911) or a local telephone call. If it is a local telephone call, the system will wait until it has sufficient digits and connect the call to a local gateway. If the next 3 digits were 911, it would connect the call to a local gateway and route to the emergency services number.

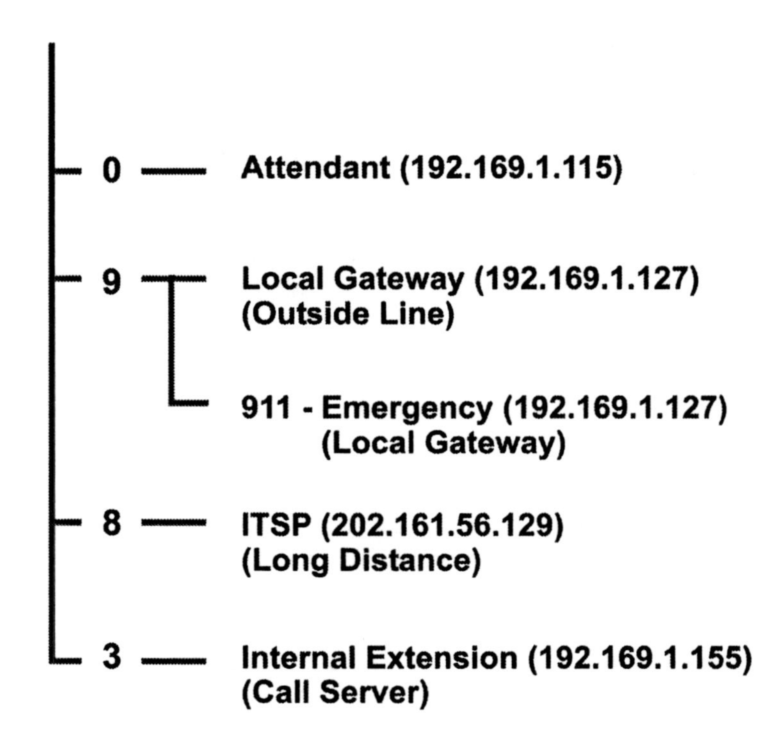

Figure 6.14, Dial Map Operation

Direct Inward Dialing (DID) Assignments

Direct Inward Dialing (DID) assignments, map incoming phone numbers to specific extensions or groups within the communication system. Examples of a DID assignments include assigning the main phone line (main telephone number) to attendant's console or to the auto attendant and assigning the fax line (fax telephone number) to the fax machine extension.

The telephone number is stored in the device's MIB file. The assignment of these telephone numbers allows for the automatic routing of calls from phones within the network and from the public telephone network to the specific devices (extensions) they are trying to reach.

Figure 6.15 shows how telephone the administrator may assign telephone numbers to SIP communication units. This diagram shows that each device that can be assigned a telephone number has a MAC address and potentially a port number.

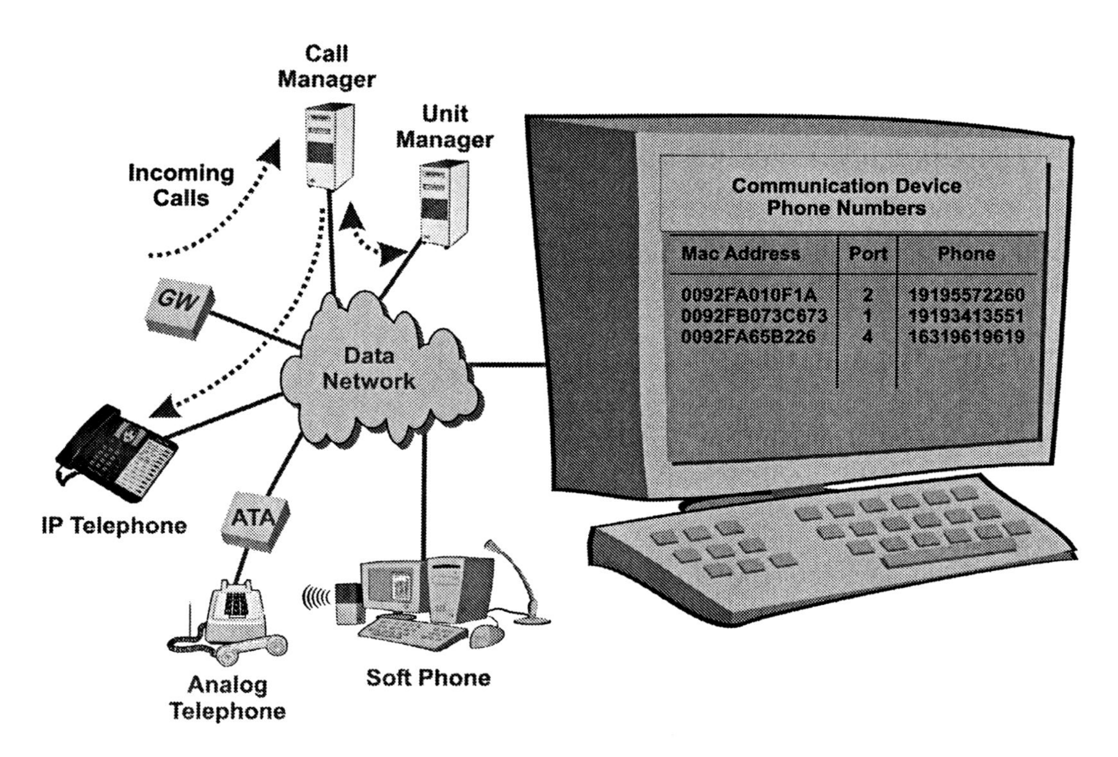

Figure 6.15, SIP Communication Unit Phone Number Assignment

Hunt Groups

A hunt group is a list of telephone numbers that are candidates for use in the delivery of an incoming call. When any of the numbers of the hunt group are called, the telephone network sequentially searches through the hunt group list to find an inactive (idle) line. When the system finds an idle line, the line will be alerted (ringing) of the incoming call. Hunt groups are sometimes called rollover lines.

Attendant Switchboard Console

An attendant is a person who answers, screens, or directs calls in a communication system. An attendant switchboard is a communication device (phone and display) that provides the ability for a receptionist (attendant) to identify and answer incoming calls, interact with callers, and redirect (transfer) calls to the proper extension. Attendant switchboard consoles in an IP PBX system are software programs that typically operate on a standard multimedia computer. Attendant switchboards display incoming Caller ID information, have graphical call status indications (hold, in-use), allow quick access to company directories, and permit the simple transfer of calls through the use double-clicks.

Automatic Call Distribution (ACD)

Automatic Call Distribution (ACD) is a system that automatically distributes incoming telephone to specific telephone sets or stations calls based on the characteristics of the call. These characteristics can include an incoming phone number or options selected by a caller using an interactive voice response (IVR) system. ACD is the process of management and control of incoming calls so that the calls are distributed evenly to attendant positions. Calls are served in the approximate order of their arrival and are routed to service positions as positions become available for handling calls.

Figure 6.16 shows a sample automatic call distribution (ACD) system that uses an interactive voice response (IVR) system to determine call routing. When an incoming call is initially received, the ACD system coordinates with the IVR system to determine the customer's selection. The ACD system then looks into the databases to retrieve the customers' account or other relevant information and transfer the call through the IP PBX to a qualified customer service representative (CSR). This diagram also shows that the ACD system may also transfer customer or related product information to the CSR.

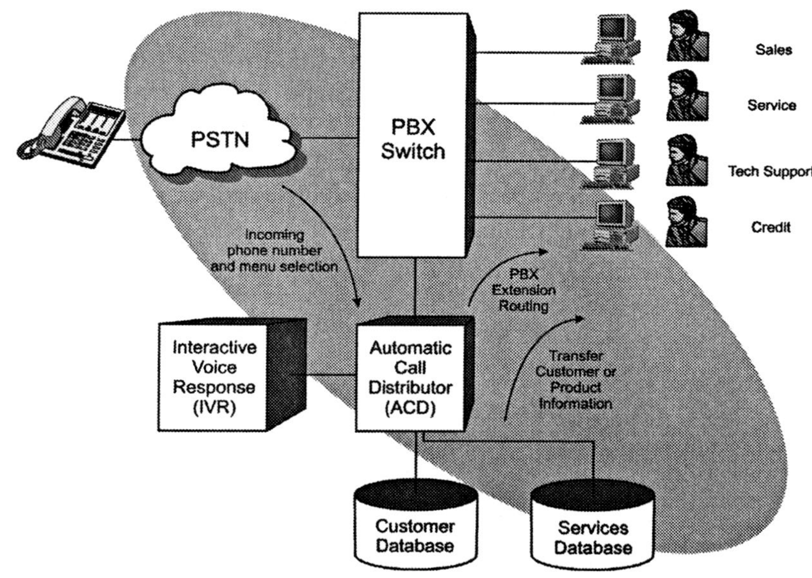

Figure 6.16, Automatic Call Distribution

Voice Mail (VM)

Voice mail is a service that provides a telephone customer with an electronic storage mailbox that can answer incoming calls and store voice messages. Voice mail systems use interactive voice response (IVR) technology to prompt callers and customers through the options available from voice mailbox systems. Voice mail systems offer advanced features not available from standard answering machines including message forwarding to other mailboxes, time of day recording and routing, special announcements and other features.

The administrator of the voice mail systems can setup and delete voice mailbox accounts for users. The users are provided with the tools to change their greeting, passwords, message forwarding, and other options.

Some of the available options for SIP based voice mail systems include message waiting notification and forwarding voice mail to email address. Voice mail notification can be sent to email address or paging addresses. The voice mail contents can be sent as an attachment to any email address allowing the user to retrieve their messages via any computer terminal that has audio capability and is connected to the Internet.

Figure 6.17 shows a voice mail server screen that allows a SIP system manager to setup new media accounts on a SIP voice mail system. This diagram shows that this web portal allows the account manager to setup new voice mailboxes, add music on hold accounts, and links to other media sources.

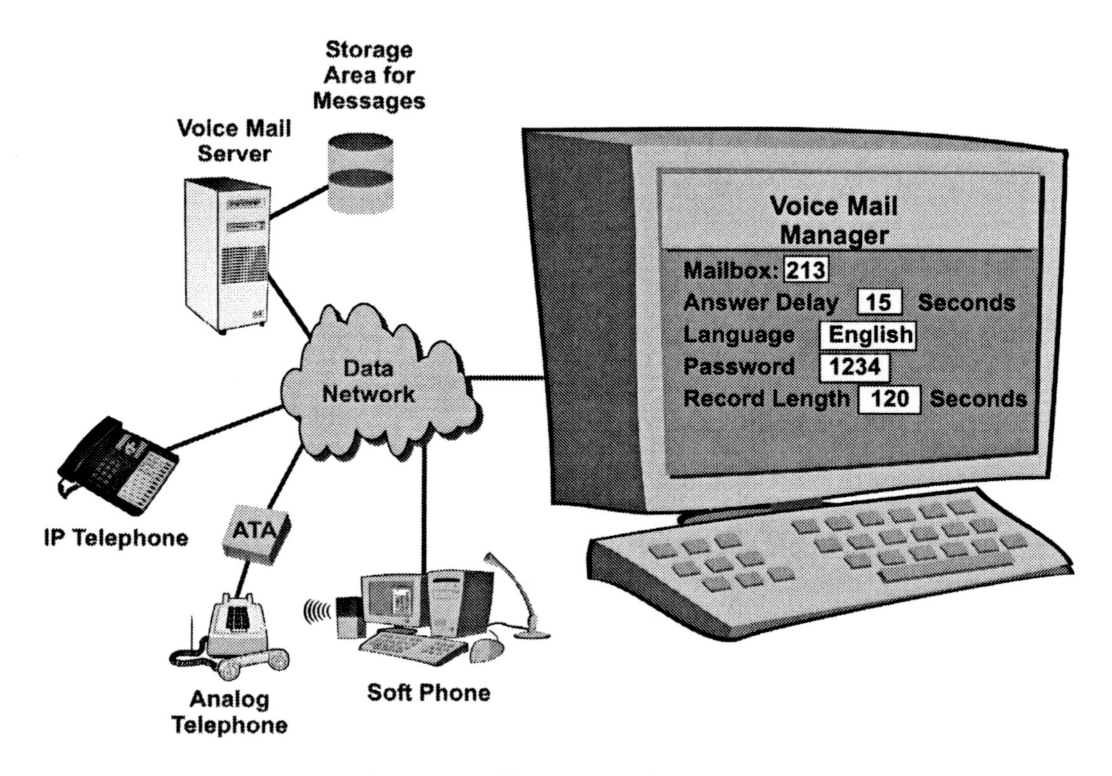

Figure 6.17, SIP Voice Mail Software

Chapter 7

Managing SIP Systems

Managing SIP systems involves service provisioning, the configuration of equipment, and the testing and maintenance of a network. For small simple systems that use hosted services (such as IP Centrex or ITSPs), the system management may simply involve the user logging into web portal to add or remove features and communication servers. For large complex systems that share data networks and have multiple wide area connection lines, the management of the system can involve the regular evaluation of system capacity and performance, setting up continuous testing (heartbeat testing), using a policy server to give priority to specific types of users and services, and the updating and maintaining of firewalls.

Service Provisioning

Service provisioning is where an authorized agent processes and submits the necessary information to enable the activation of a service. This typically includes the setup of; transmission, wiring, equipment configuration, account creation, and service activation.

Figure 7.1 shows how an administrator can provision services in a SIP system. In this example, a system administrator receives a request to setup a new user. This example shows that the user (or their manager) has completed an online form indicating a new account is to be setup and the type of equipment that will be used. The system administrator logs onto the system administration portal and creates a new account in the system (new account ID). This account ID code is used to associate all of the features and services with the device(s) assigned to the account user.

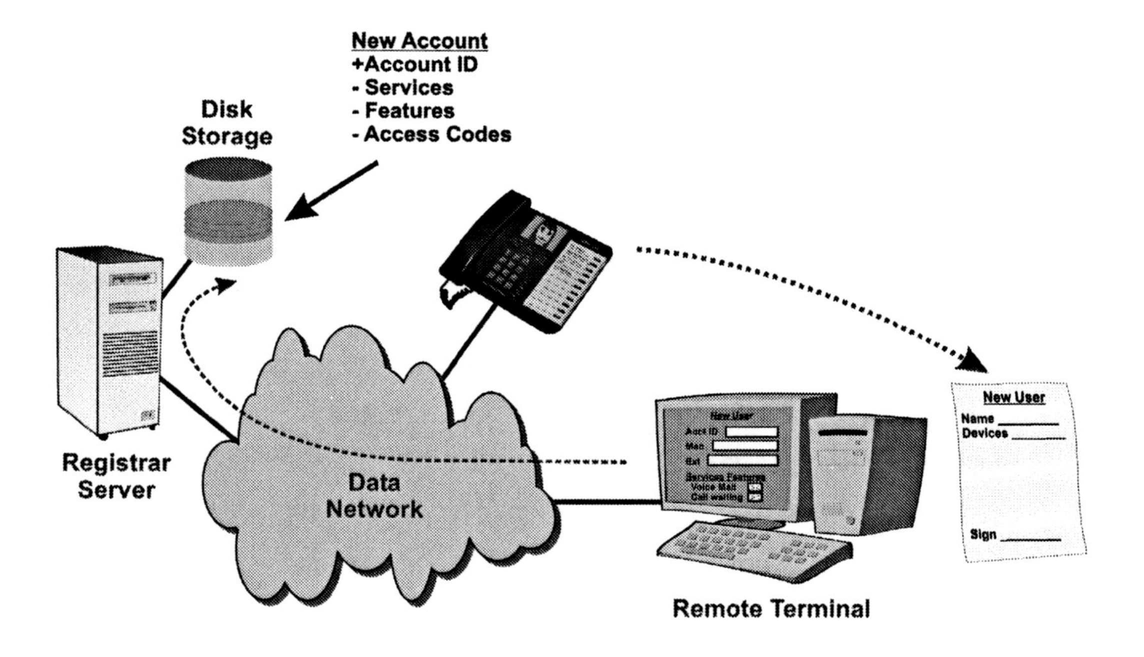

Figure 7.1, SIP Service Provisioning

System Capacity Allocation

The system capacity is the amount of resources (such as bandwidth and call processing) that is allocated for different types of services. The system capacity allocation can be automatic or it can be specifically assigned for different classes of services and configurations.

When SIP voice systems use the same data connections as computer networks, these systems must share the available bandwidth. The relative bandwidth for each call is generally less than 0.1% of the capacity of the 100 BaseT data communication lines used by most companies. Because SIP-based communication systems are usually implemented on Ethernet systems, an amount of bandwidth is automatically assigned (allocated) for each shared communication session (such as voice and data). This is a result of the fundamental design of the Ethernet communication system. Each device in the Ethernet system competes to transmit data. After a device has competed and transmitted a packet of data, its access priority is decreased. This allows other devices to have priority when accessing the system. Because the Ethernet system allows a maximum packet data size of 1500 bytes and each device is given fair access to the data link regardless of how much data it must send, the Ethernet system automatically allocates regular access to each device that is connected to the system.

Figure 7.2 shows how a data network shares bandwidth for both voice and data communications. This diagram shows that a single router is providing data communications service to IP telephones and computer workstations. In this example, a computer workstation is transferring a large file and the IP telephone is continuously sending a small amount of data (90 kbps). Because the LAN data network (Ethernet) has a maximum packet size of 1500 bytes of data and a standard high-speed data transmission rate of 100 Mbps, the router automatically divides the large file into smaller data blocks and access is shared between the IP telephone and the computer workstation. When the data packets arrive at the relatively low-speed WAN connection, congestion can occur. If congestion were to occur, the router connected to the WAN connection would begin to delay the transmission of packets. In this example, the WAN router gives priority to the voice over data network packets and delays the file transfer packets.

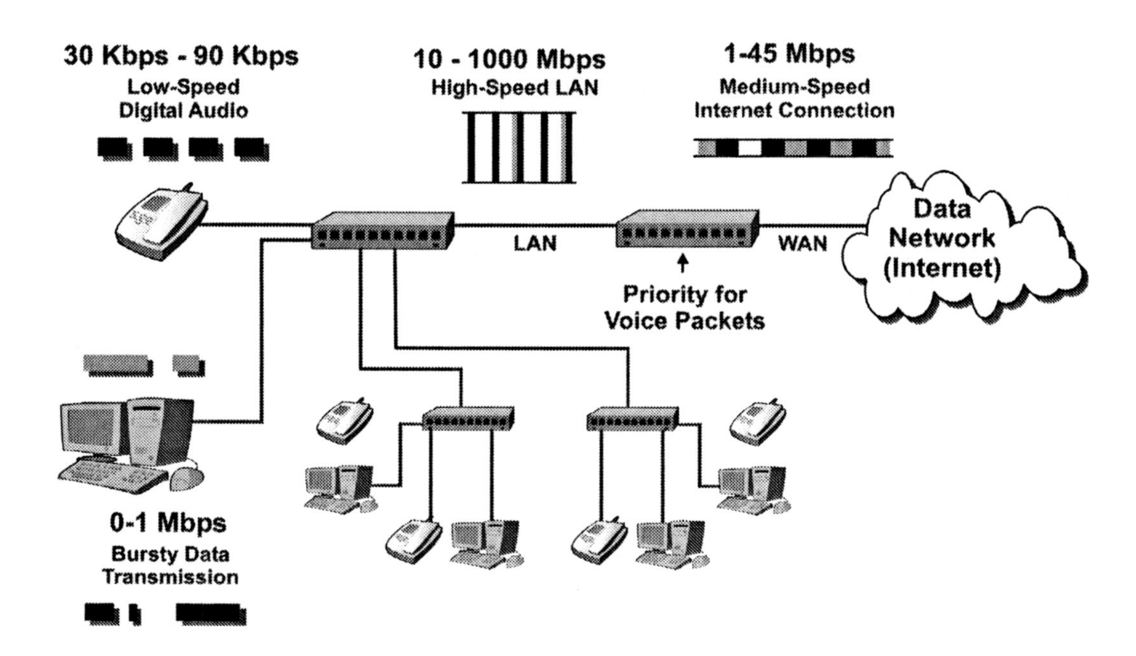

Figure 7.2, Sharing Bandwidth

While the automatic allocation of bandwidth may work fine for small systems with a limited number of devices, as the number of devices and switching points (routers) increases, capacity evaluation and adjustment should be performed.

Figure 7.3 shows how a data network may be monitored for several days to determine the capacity and transmission delays at concentration points (routers and switches) within the network. In this example, a data network has several routers that transfer data between computers in the company and to other computers connected to the Internet. As part of the VoIP capability pre-test, each router is monitored for peak data transfer activity for several days. This helps to determine if the data network lines and switching points have enough capacity to provide both the data network and VoIP system needs. In this system, an on-site company host computer stores the files for engineering and sales. An off-site company computer is used to store files from the company computer that are transferred through the Internet during the late evening hours. The analysis shows that the engineering router uses bandwidth during the morning and evening hours (light lunchtime use). The sales router has capacity used from late morning through late evening (they eat lunch at the office). The analysis also shows that the company's Internet data connection has high capacity late in the evening when information backup is in process.

It is perfectly acceptable to apply traditional teletraffic engineering theories to calculate whether an IP-based network has sufficient bandwidth to support IP telephony and operate with the required Quality of Service (QoS). By assessing the number of users at a site and their typical calling patterns, (number and duration of calls), it is possible using standard formulas to calculate the total telephony traffic generated. The total bandwidth requirements can then be calculated from this traffic volume by accounting for the codecs being used and any overhead requirements from the transmission protocols. Standard traffic theories can then be applied to assess whether the network has sufficient available bandwidth to support the IP telephony with a satisfactory level of blocking, or Grade of Service (GoS).

In addition to ensuring that the network and the network components have sufficient capacity to handle the additional traffic that arises from telephony, an assessment of the network performance should also be conducted. Various analyzers and software tools are available that will allow network performance to be measured. The performance parameters include; timing jitter (variations in the delay experienced by each packet in a stream), lost packets, out-of-sequence packets and overall latency. It may be necessary to connect a call simulator to the network to perform a 'stress test' in which the network is bombarded with call attempts to assess it's performance during periods of peak traffic. Importantly this testing should extend to the gateway devices to ensure that the volumes of traffic generated on the 'packet-side' of the network can be handled by the gateway and efficiently passed to the 'circuit-side' of the network.

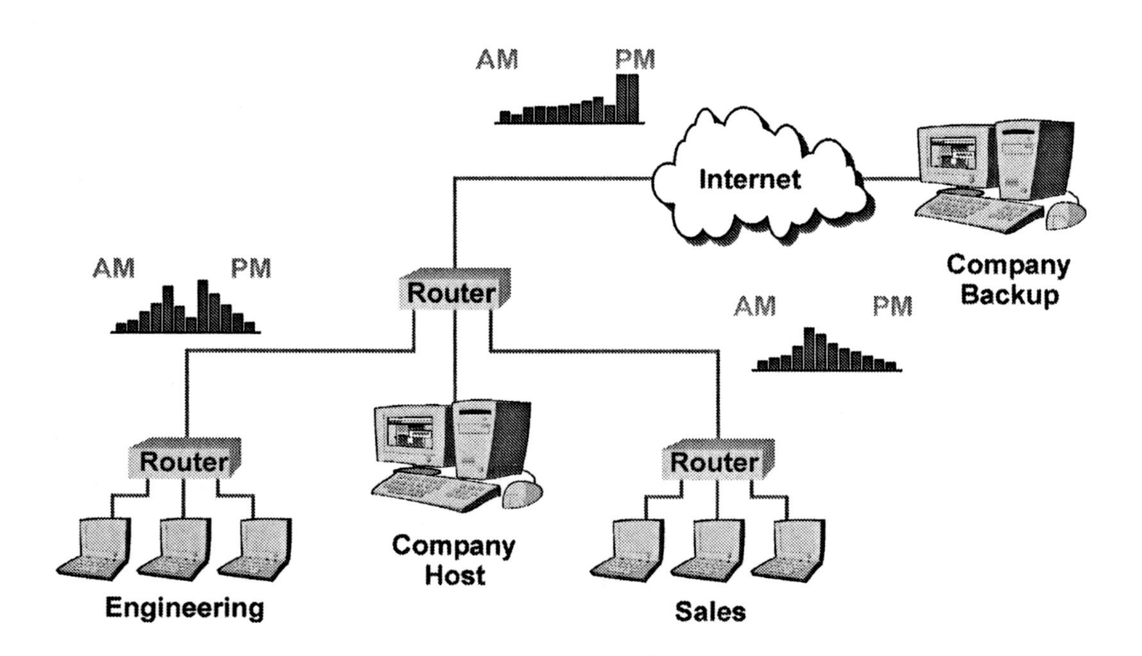

Figure 7.3, Data Network Capacity Evaluation

Policy Server

A policy server is a communications server (computer with a software application) that coordinates the allocation of network resources based on predetermined policies such as the priorities and resources required by communication services and applications within the network. A policy server is used to help to provide different levels of quality of service (QoS) and to manage network operation in the event of loss of resources (prioritize services). The policy server uses pre-set policies to define which communication services are critical (such as voice) and how much resource should be allocated to these critical services at the expense of other communication services (such as web browsing).

Quality of service (QoS) is one or more measurements of the desired performance and priorities of a communications system. QoS measurements may include service availability, maximum Bit Error Rate (BER), minimum Committed Bit Rate (CBR) and other parameters that are used to ensure quality communications service. A policy server may assign, or prioritize, resources or bandwidth. This can be used to ensure specific services (such as voice) are given priority over other services (such as data).

Bandwidth reservation is a process that is used to reserve bandwidth capacity through devices or communications lines for specific sessions or services. Bandwidth reservation is commonly performed through the use of reservation protocol (RSVP). RSVP is used to reserve an amount of bandwidth dedicated to packet flow for specific communication sessions in a packet data network. RSVP is primarily used in real-time communication sessions (such as voice over packet).

Additional IP telephony protocols include Common Open Policy Service (COPS) and the Open Settlement Protocol (OSP). COPS are a protocol that allows a system to implement policy decisions by allowing a client to obtain system configuration and parameter information from a policy server. OSP is a standard protocol that is designed to transfer billing information to allow inter-carrier billing between voice and data communication systems. The OSP format is approved by the European Telecommunications Standards Institute (ETSI). OSP allows communication gateways to transfer call routing and accounting information to clearinghouses for account settlement between carriers (service providers), which ultimately allows ITSPs to interwork their networks.

Figure 7.4 shows how a policy server can be used in a SIP system to assign priorities to specific users and services. This example shows that a policy server has detected that a wide area communication link has been disabled. This policy server has been programmed to automatically reallocate and reserve bandwidth on an alternate router for voice based services.

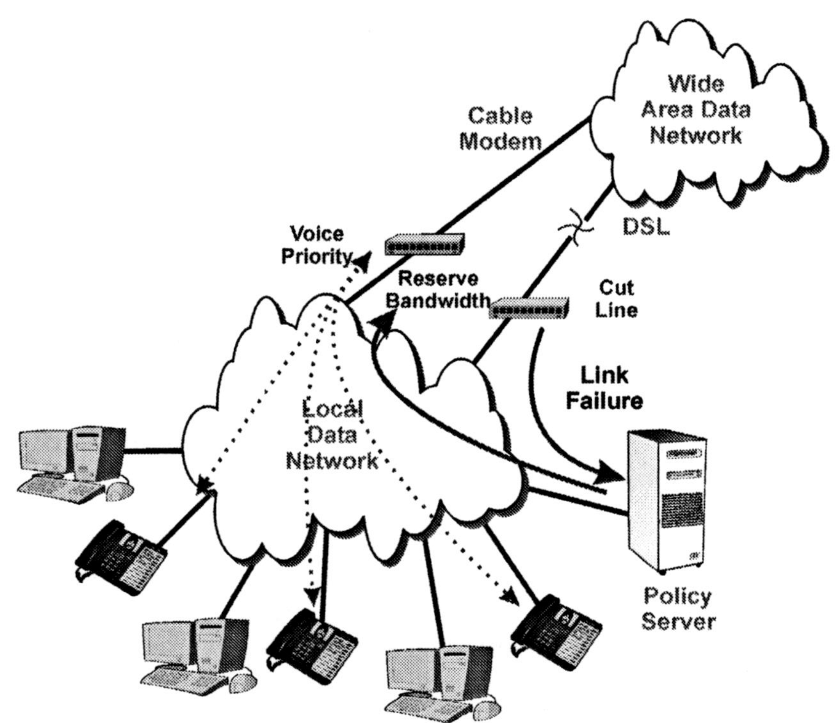

Figure 7.4, SIP System Policy Server

Equipment Configuration

Equipment configuration is the process of sending information to a device that is used to adapt the equipment or a software program to its environment (configuration). SIP system equipment can be manually configured (via the device or a remote portal) or a system administrator can configure it by the transfer of information.

Management Information Bases (MIBs) are a collection of definitions, which describe the properties of the managed object within the device to be managed. Every managed device keeps a database of values for each of the definitions written in the MIB. MIBs are used in conjunction with the simple network management protocol (SNMP) as well as Remote Monitoring (RMON) to manage networks. MIBs (referred to now as MIB-I) were originally defined in RFC1066.

Trivial File Transfer Protocol (TFTP) is a protocol that is used to transfer files between devices in a data communication network. TFTP is a simplified version of File Transfer Protocol (FTP) and it is commonly used in devices to allow for the transfer of setup and configuration information. TFTP is defined in RFC 1350.

Figure 7.5 shows how a user manager service can automatically configure and upgrade equipment in a SIP communication system by downloading configuration files. In this example, the configuration file that contains the server addresses and preferred features is transferred to the SIP device and is stored within the MIB memory is of the device.

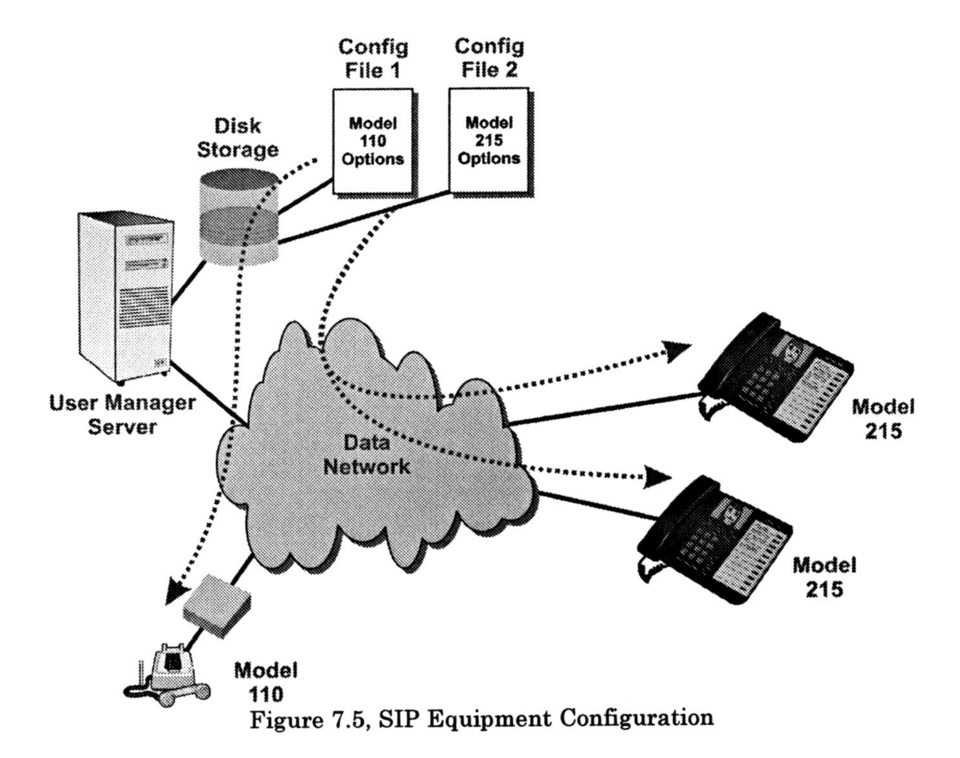

Figure 7.5, SIP Equipment Configuration

Equipment Firmware

Firmware is software program instructions that are stored in a hardware device that performs data manipulation (e.g. device operation) and signal processing (e.g. signal modulation and filtering) functions. Firmware is stored in memory chips that may or may not be changeable after the product is manufactured. In some cases, firmware may be upgraded after the product is produced to allow performance improvements or to fix operational difficulties.

Some device manufacturers provide an Internet site that contains the latest version of firmware for their devices. Firmware may be downloaded from these sites using TFTP.

Figure 7.6 shows a simple process used to upgrade the operating software (firmware) of a device within a SIP system. This example shows how a SIP telephone that is equipped with flash memory can receive new software (firmware). This example shows that a section of incorrect code (a bug) in the original software could be repaired. A portion of the operating software manages this transfer of firmware; that cannot be changed or updated. During this transfer and the updating process, the user cannot control the device.

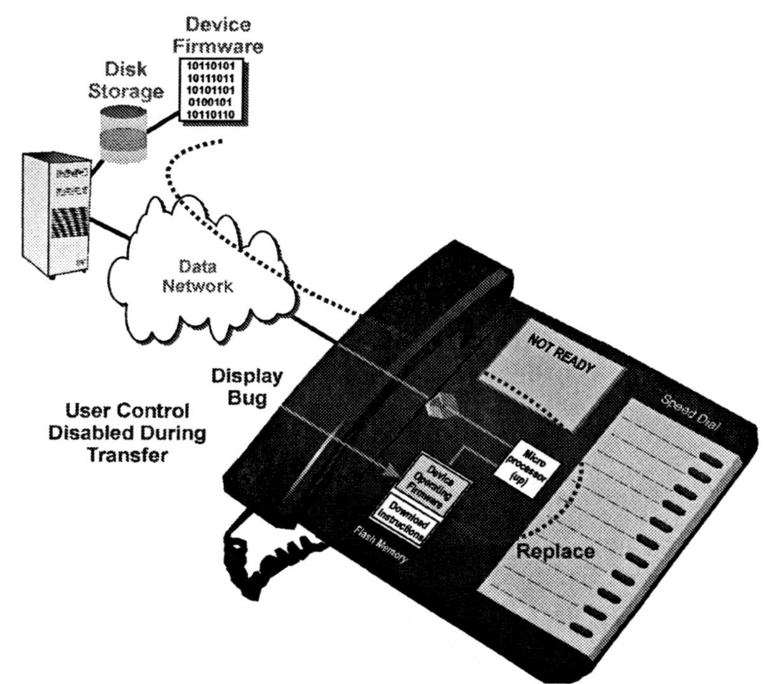

Figure 7.6, SIP Firmware Upgrade

Testing and Maintenance

Testing and maintenance are the processes used to ensure the acceptable operation of equipment and services for users or a system. Testing may be automatic, continuous, or diagnostic.

The Simple Network Management Protocol (SNMP) is a standard protocol used to communicate management information between the Network Management Stations (NMS) and the agents (such as routers, switches, and network devices) in the network elements. By conforming to this protocol, equipment assemblies that are produced by different manufacturers can be managed by a single program. SNMP protocol is widely used in Internet protocol (IP) environments and operates over the UDP well-known ports of 161 and 162. SNMP was originally defined in RFC1098 that is now obsolete and updated by RFC1157.

The SNMP protocol is intentionally simple in it's structure and contains only a limited number of message types. These messages allow the NMS to view and set parameters in a managed object and for the object to report parameter values back to the NMS when requested. Management data in an SNMP network can also be collected via a mechanism known as a trap. When a predefined trap is triggered in an SNMP agent, like a network device, it will asynchronously deliver an event notification to a network management station (NMS) or trap receiver. SNMP traps are defined in MIB definitions by setting threshold values which when exceeded will trigger a trap message. SNMP traps are sent on a best-effort basis and without any method to verify whether or not they were received by the trap receiver.

When information such as events or measurements is created in a network, they are commonly stored in a log file. The log file is continually updated (added to) as new events occur. Log files are used to analyze problems that have occurred, or may occur, within a particular application or service.

Figure 7.7 shows how a log file gathers and stores information from various parts of the network. This example shows that a system server maintains a log file and that this log file stores the events (alerts), time of the event, and other relevant data associated with the event. The first event in this diagram occurs when an IP telephone device is connected to a data socket at 0907. The next event is recorded at 1015 that indicates a file was lost during transfer due to a lightning noise spike. The final event in this example is the recording of an open connection (cut wire) to one of the data terminals.

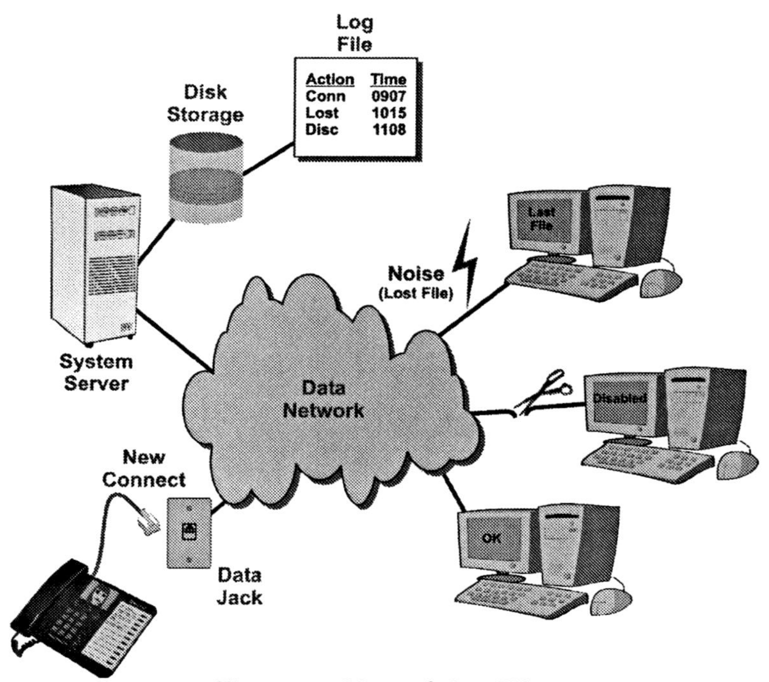

Figure 7.7, Network Log Files

An additional management feature may be provided by means of a device known as Heartbeat server. Heartbeat servers can transmit continuous signals called heartbeats that are transferred through critical parts of the network. Because these regular heartbeats are expected to be received at different parts of the network, missing or delayed heartbeats can be used to determine if there is something wrong with the system. A heartbeat sometimes may be used to automatically reconfigure or restart equipment in the event of an equipment failure. There is normally only one heartbeat server on a network or system.

Figure 7.8 shows how heartbeat testing can be used in a SIP system to ensure the system is operating correctly. This diagram shows a heartbeat server is sending out continuous test packets that are routed through different types of servers. This example shows that heartbeat test packets are sent along with standard packets through the data network. This system has been setup to forward heartbeats through the proxy server (call manager), through the gateway server, through the feature server, and back to the heartbeat server. If any of these packets are missing or delayed, this would indicate that one of these 3 servers is experiencing a problem and action can be taken immediately.

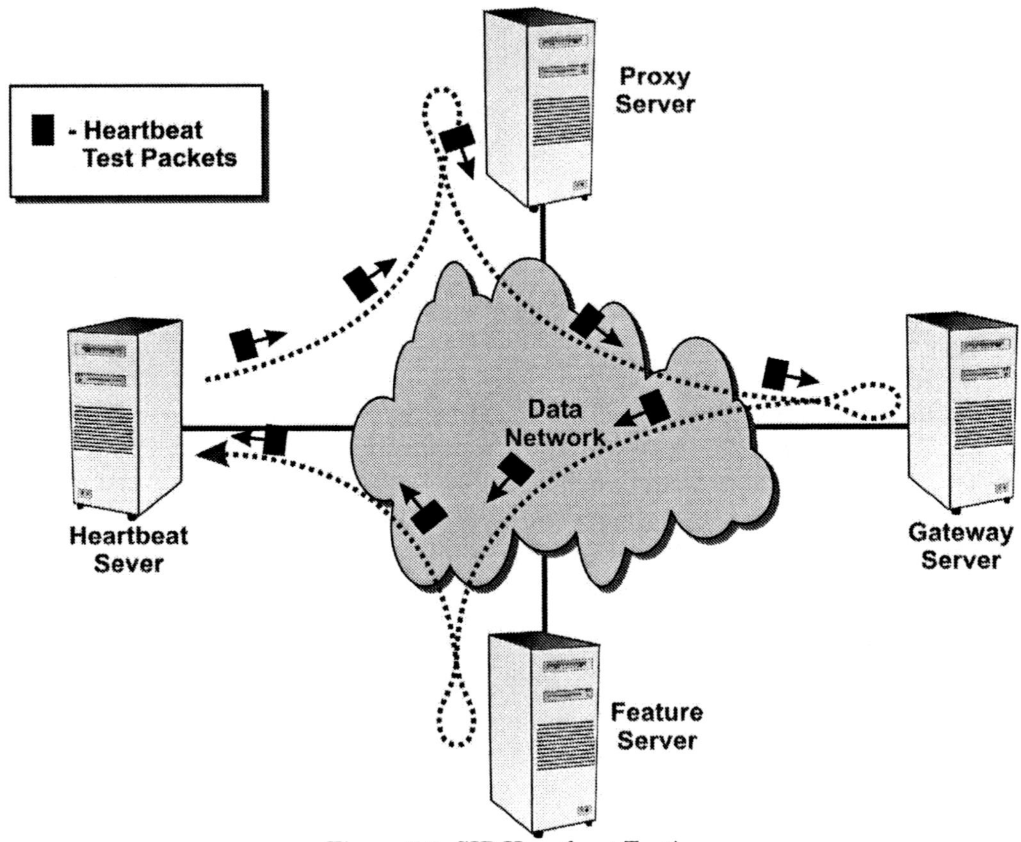

Figure 7.8, SIP Heartbeat Testing

When equipment fails in a system, testing is performed to identify and isolate the failed equipment or sub-system. A common form of testing used in systems is loopback testing. Loopback testing is a process of testing the transmission capability and functioning of equipment within a system in which a signal is transmitted through a loop that returns the signal to its source. The test verifies the capability of the source to transmit and receive signals.

Figure 7.9 shows how loopback testing can be used in a SIP system to progressively test, confirm, and identify failed equipments or portions of a network. In this example, the test signals is created by a test device that is connected to a local area data network. This example shows that the first test involves programming the media gateway to loopback mode so the received test signal from the test device can be returned to the test device. The test device can report if the signal was received and what the quality of the signal is (how many errors). The second test involves programming a remote gateway to loopback mode. This test confirms that the local data network, local media gateway, and wide area network are functioning correctly. The third test in this example sets a remote test device to loopback mode. This test confirms that the local data network, local media gateway, wide area network, remote media gateway, remote data network, and remote test device are working correctly. Failure of one or more of these tests can be used to isolate and help diagnose problems with the system.

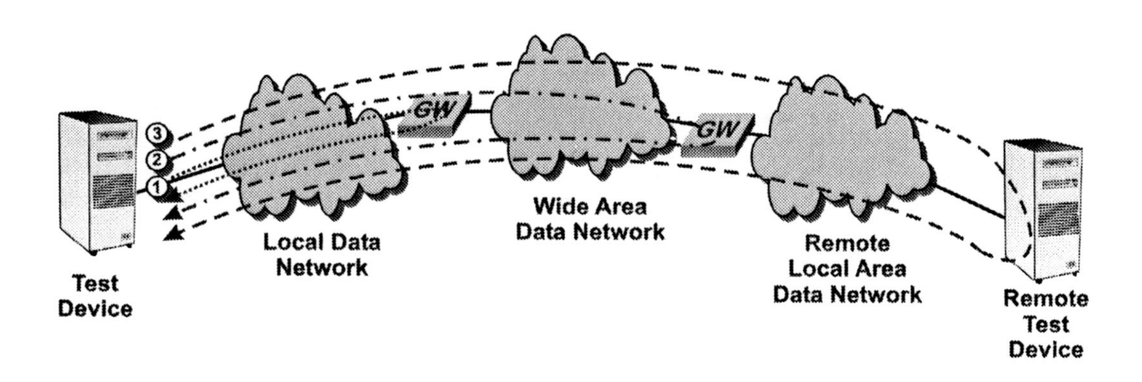

Figure 7.9, SIP System Loopback Testing

Test Plan

A test plan is used to validate the overall operation of a device or system. A test plan is sequence of tests that ensure the system will operate correctly and if the system is capable of operating in the event of one or more equipment failures.

A test plan should test all of the end points (telephones and gateways), the capacity of the system, and validate that the amount of resources used (such as processing power) are within design limits.

A test plan should also be used to determine if alternative communication paths would automatically be used in the event of a detected failure. The test plan will usually specify the insertion of simulated (or real) faults to determine if the system will automatically recover in the event of failures.

Because different call states during different feature operation (such as voice mail and call waiting) may cause different responses to service requests and messages, the test plan should specify the features that should be tested and in which mode of operation the particular feature should be set during the test.

Due to the variable timing of commands (such as a gateway waiting for a dial tone), erratic operation can result. The test plan should repeatedly test equipment that can have different results due to variable timing.

The test plan should ensure different types of devices, or devices produced by different manufacturers, should be tested with each other. While the industry standards were created to ensure reliable operation, different manufacturers (and sometimes different designers within the same manufacturer) may interpret and implement the industry standards slightly differently. This can cause erratic operation.

Figure 7.10 shows a test plan that may be used for a SIP system. This figure shows how a test plan is essentially a list of tests that progressively certify that a system or network is performing as expected. In this figure, the first test group involves testing the current system capacity and reliability.

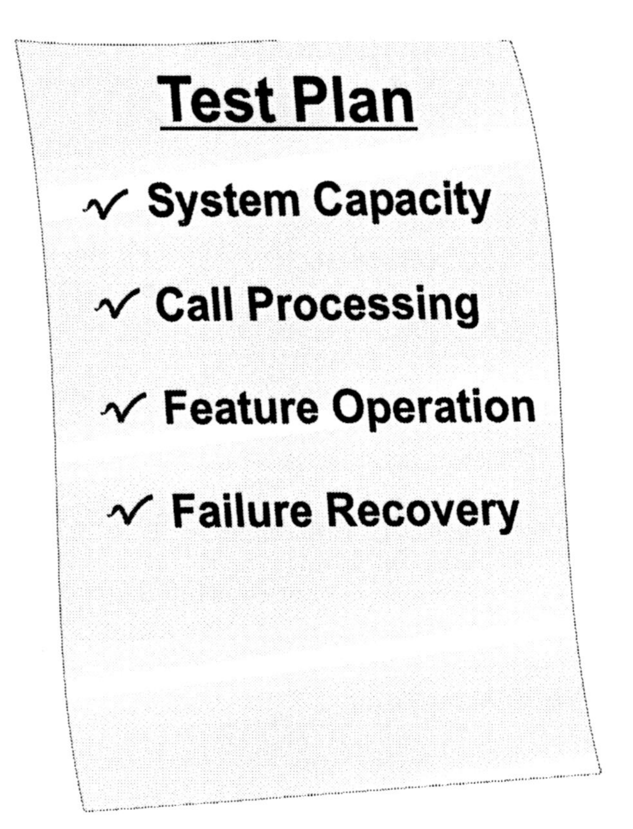

Figure 7.10, SIP Test Plan

This is a list of Key Tests in a test plan, in logical order

1. System Capacity Test (data, voice)
2. Call processing
3. Feature Operation
4. Failure Recovery

Network Address Translation (NAT)

Network address translation (NAT) is a process that converts network addresses between two different networks. NAT is typically used to convert public network addresses (such as IP addresses) into private local network addresses that are not recognized on the Internet. NAT provides added security, as computers connected through public networks cannot access local computers with private network addresses.

NAT translation may also involve the changing of port numbers. Applications share the same IP address by assigning communication channels to port numbers. Some port numbers are pre-assigned for specific types of application (such as FTP, Email, and SIP). The changing of these port numbers can result in the wrong application protocol being used and/or a loss of a communication channel.

To get through a NAT and allow two-way communication, stay-alive protocols are used. These protocols regularly transmit small amounts of data to keep a communication session alive. This allows a communication channel to remain open with the server so incoming calls can be received.

Figure 7.11 shows the basic operation of network address translation (NAT) system. In this diagram, the NAT receives a message with a desired pubic IP address (209.67.22.59) originating from a local computer with a private IP address of 10.1.1.1. The NAT translates this originating address to a public IP originating address. The NAT then initiates as session with the Internet server (web site) using the network's public IP address 118.54.23.11 as the originating address. The Internet server receives the request for information and responds with data messages address to the NAT's public IP address. When the NAT receives these data messages for that particular communications session, they are translated to the local (private) IP address 10.1.1.1 and forwarded to the originating computer. If messages are received to the NAT's public IP address that are not part of a communications session that it knows about, the NAT will not route the messages to computers connected to the LAN.

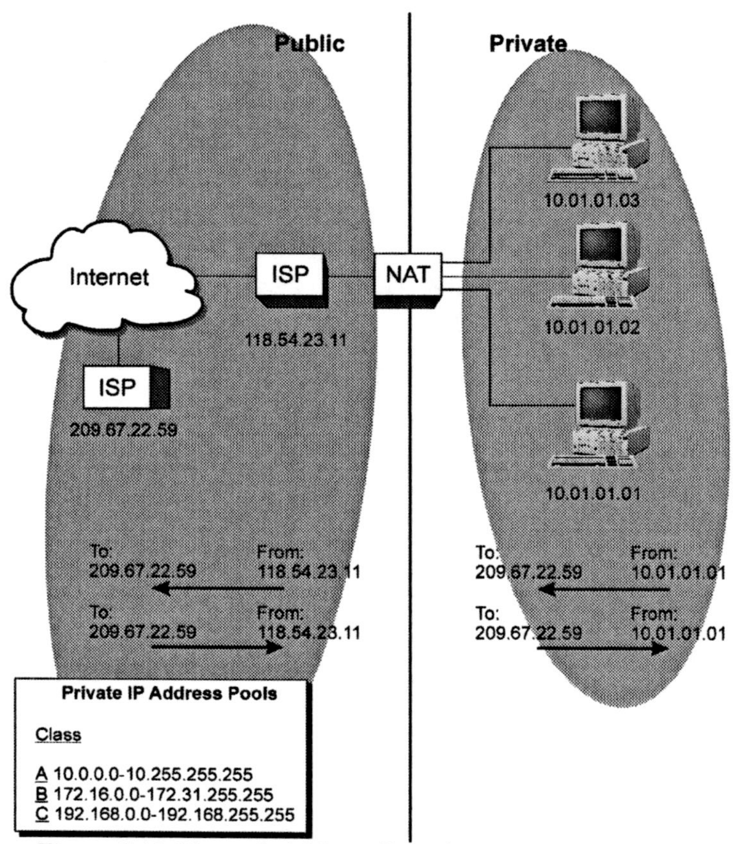

Figure 7.11, Network Address Translation (NAT) Operation

Firewalls

A firewall is a data-filtering device that is installed between a computer server or data communication device and a public network (e.g. the Internet). A firewall continuously looks for data patterns that indicate unauthorized use or unwanted communications to the server. Firewalls vary in the amount of buffering and filtering they are capable of providing. An ideal (perfect) firewall is called a "brick wall firewall."

Getting through the firewall is called "Firewall Traversal." Firewalls create small data transmission paths between the inside network (protected area) and other networks (such as the Internet).

Firewalls can cause considerable problems for real time communication systems such as SIP. Firewalls add delays and they may block certain types of protocols. To overcome the problem of firewalls, there are several options available for system designers. These include installing the voice communication system ahead of existing firewalls (bypassing the firewall), updating firewalls to detect and allow SIP or other VoIP protocols, or installing Application Level Gateway (ALG) firewalls that can detect and modify IP telephony packets that enter and leave the system.

Figure 7.12 shows some of the firewall options that are available for SIP systems. Option 1 shows how the SIP system is installed ahead of the firewall that protects the company's information system. Option 2 shows a system that has upgraded the firewall to allow SIP protocols. Option 3 shows a system that has installed an ALG firewall that can detect and modify packets that are transmitted between the external (public) system and internal (private secure) system.

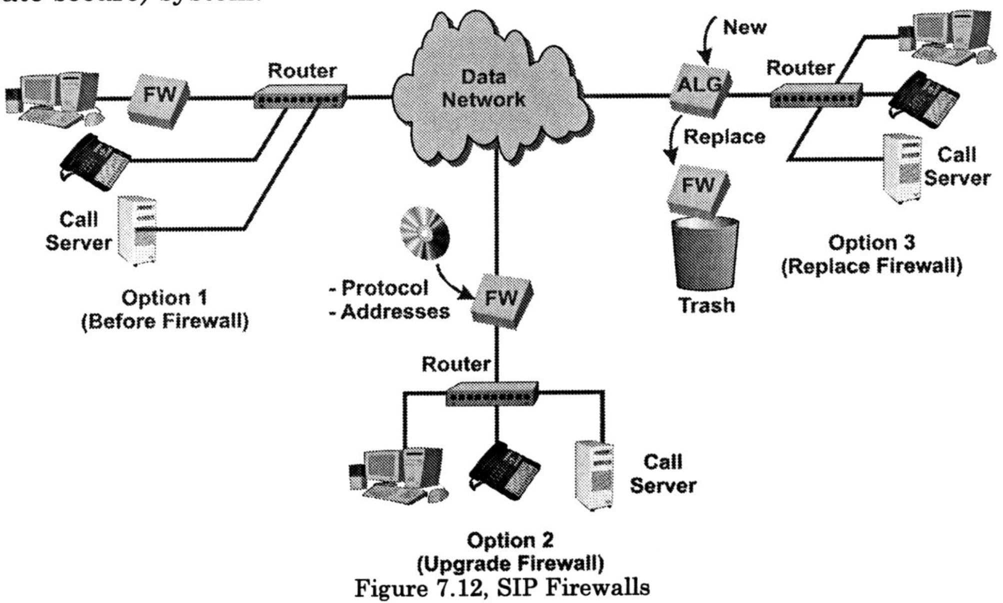

Figure 7.12, SIP Firewalls

Integrated System Management

As with a traditional PBX-based telephony system there is a requirement to monitor the performance of an IP telephony system in terms of call quality, call success rates, bandwidth and trunk utilization, and system status. Most major vendors of IP telephony equipment now provide integrated configuration and management tools that allow system administrators to configure the IP system, including individual phones, and to manage the system from a web-based interface. The configuration element of the management software will allow the administrator to setup accounts, set IP routing and Quality of Service (QoS) parameters, configure trunks and perform the necessary moves adds and changes on the system.

When monitoring the network the management tool should allow the administrator to view the status of calls in progress, including parameters such as the calling and called parties, view trunk utilization and the IP bandwidth allocated to calls.

Many of the tools provided support SNMP and will allow the administrator to set the traps and their thresholds and to view alarms and events within the system. Typically at the web interface the system administrator should be able to see when various events such have occurred, such as calls exceeding a threshold or the QoS requirement not being satisfied. Event and alarm logs will be provided so that analysis can be performed by alarm/event type, time or severity.

Management software should also be capable of providing reports on system activity, alarms and events. In some cases vendors also provide trend reporting capabilities that permit system administrators to analyze call history and trends.

Chapter 8

SIP System Economics

SIP system economics involve equipment costs, software costs, service usage costs, and other less obvious (indirect or hidden) system costs. The selection and use of SIP systems typically results in cost reductions, productivity improvements, and new revenue producing applications.

At the end of 2003, the typical cost per SIP telephone station ranges from approximately $75 to $1,000 compared to $125 to $1200 for a traditional (legacy) digital PBX station [ ]. The usage toll cost (if any) for IP Telephony ranges from $0 per minute (data to data call) to 4 cents per minute for domestic calls and 3 cents per minute to 10 cents for most international calls compared to 9 cents per minute for domestic long distance calls in the United States and more than 47 cents per minute for international calls when provided by traditional telephone companies [[i]].

Equipment Cost

Basic SIP system equipment costs include the cost of call servers (computers), communication lines, data network routers, backup power supplies, and IP telephones or analog telephone adapters. Cost reductions that result from the use of SIP communication equipment results from the availability of low cost standard equipment.

Call servers are computers that run a network operating system such as Windows or Linux based servers. Computer servers are usually reliable computers with redundant assemblies such as multiple processors, Redundant Array of Independent Disks (RAID), or Storage Area Network (SAN) information storage systems. For IP Telephony systems, there is usually at least one server per company location. Because the IP Telephony call server is only responsible for the setup and management of the call and not the media delivery system, the call server may be a relatively low specification computer system.

For IP Telephony systems that use the same data network cabling and infrastructure for computer terminals and IP telephones, there may be small or no additional cost for the interconnecting data network equipment used by IP telephones. However, some data network equipment may need to be added to increase system reliability (duplicate paths) and the capacity of data transmission lines may be increased between specific locations.

Gateways convert the media from SIP communications systems into formats suitable for other networks such as the Public Switched Telephone Network (PSTN). The cost of gateways can range from approximately $200 for a single line gateway to more than $5,000 for a multi-line trunking gateway. SIP systems commonly use gateways to connect a SIP system to local telephone lines. Optionally, gateways may be installed and used in remote locations to allow toll bypass for the connection of calls to distant international locations.

IP telephones can be softphones, analog telephone adapters (ATAs), or hardphones (IP Telephones). The cost of IP telephones can vary from no cost for a softphone that is provided free and installed on a multimedia computer, to over $1,000 for a full featured IP telephone.

Reductions in equipment costs result from the use of low-cost standard data communication equipment (routers and hubs) instead of proprietary private telephone switches and PBX telephones.

Figure 8.1 shows some typical equipment costs associated with the use of an IP PBX system that is added to an existing data network. This cost structure assumes that the data network is capable of providing both data and voice services within the existing bandwidth. This table shows that a low-cost computer call server is used. The cost of purchasing gateways can range from $200 each (single line) to $5,000 each (multiple line trunking gateway). The cost of IP telephones can range from zero for softphones installed on existing multimedia computers to more than $1,000 for advanced IP Telephone devices.

Item	Cost
Call Server(s)	$500 to $3,000
Gateway(s) to PSTN	$200 to $5,000
IP Telephones	
Softphone	$0 to $200
Analog telephone adapter	$50 to $300
IP Telephone	$75 to $1000

Figure 8.1, SIP Equipment Cost

Software Cost

SIP system software costs include the cost of server operating systems, server application software, optional feature software, application development tools, and software license fees. Cost reductions that comes from the use of SIP deployment result from the use of low-cost operating systems and SIP-based software, instead of typically expensive proprietary PBX software.

Popular operating systems include Windows NT, Windows 2000, or Linux. The cost of these operating systems range from $0 (e.g. Linux shareware) to over $5,000 for an operating system that has a license that allows many users to connect to the server [ ].

Server call processing software includes the call management, user management, configuration management, and administrator servers. The cost of the server software packages ranges from $0 (such as shareware from Vovida) to over $200 per user for commercial software [ ].

Network management software is used for network monitoring, quality of service (QoS) enforcement, and diagnostics. Small off-the-shelf systems with only a few users may have little need for network management software, in a large-scale deployments however, there may be a requirement for custom software that monitors and adjusts specific system elements and costs several hundred thousand dollars.

Additional software may be required in a SIP system to provide advanced features such as voice mail, automatic call distribution, web server integration, and inventory management systems. The cost for additional SIP-based call software can range from a few hundred dollars for a prepackaged system such as "Call Center in a Box" to thousands of dollars per feature for custom software. The cost of advanced features (such as advanced call center software) can total several thousand dollars per user.

Optionally, custom software development tools may be purchased so a company can develop it's own software features. This is commonly done to allow for the integration of the IP telephone system with other programs such as order processing systems, web servers, and inventory management systems.

Software development tools can cost several thousand dollars to tens of thousands of dollars.

Software license fees include the cost of clients on the server and the use of software on the end user devices (such as a Cisco ATA). Some of the equipment may be purchased with the software license fee included (such as IP telephones with installed SIP software).

Software cost reductions come from the use of low-cost feature development tools for standard call processing software and the ability to integrate the telephone systems with information systems without the need to purchase proprietary or custom software.

Figure 8.2 shows some of the common software costs for the setup of an IP Telephony system. This table shows that the operating system cost can range from zero (for open source code such as Linux) to more than $5,000 (for operating systems such as Windows 2000 with many users). The cost for call management server software will typically vary from $50 to $200 per user based on the total number of users that will be setup with the system. The cost of network management software ranges from zero (no network management software) to more then $100,000 depending on the monitoring, testing and reporting capabilities and the size/complexity of the network. The cost of software modules for advanced features (such as voice mail or call center software) can add $10 to $50 per user for basic features.

Item	Cost
Operating System	$0 to $5,000+
Call management server	$50 to $200 per user
Network management	$0 to $100,000+
Enhanced feature modules	$10 to $50 per user
Software development application	$500 to $50,000
Software license fees	$10 to $100 per user

Figure 8.2, SIP Software Cost

Service Cost

SIP system service variable costs include gateway access fees, telephone number assignment fees, Internet and other data transmission costs, and the leasing of data communication lines. The cost reductions for SIP service usage include savings in international and domestic toll charges and intelligent network service cost reductions.

While there may be no direct usage cost for IP telephony calls completed within a data network. There is a usually a usage cost when calls pass from the IP telephone system through gateways to the public switched telephone network. These costs can either be a monthly recurring charge (MRC) and/or a usage fee from another company that owns the gateway. If the termination gateway is owned by the company originating the call, the gateway cost comes from the monthly line charge for the telephone line connected to the gateway. If another company owns the gateway, gateway access fees for call termination and call origination range from approximately 1 cent to 5 cents per minute. If the gateway has telephone numbers assigned to it (to allow IP telephony calls to be received by a public telephone number), there may be an additional charge of $5 to $40 per telephone number.

Some data transmission lines (such as Internet and frame relay connections) have data transmission usage based costs. The additional data that is transmitted by IP telephony systems may require an increase in the committed data transmission rate or it may require a change in the Quality of Service (QoS) requirements for interconnecting data transmission lines, this is to ensure voice packets are not discarded during period of congestion. If the existing data transmission lines have enough capacity (committed capacity) to service the additional IP telephony usage, there should be no increase in cost of data transmission. Assuming the average commercial telephone user talks for one hour per day (typical for many businesses) and the data transmission capacity or type of service (TOS) has to be changed to accommodate the increased data transmission requirement (higher committed data rate), the result can be an increase in the data transmission costs of up to $10 per user per month.

For IP telephony systems that have multiple systems located at distant locations, leased data lines may be installed between locations. Leased lines typically have a monthly recurring charge (MRC) based on the dedicated capacity and number of ports provided by the leased data lines. The cost of leased lines ranges from approximately $40 to more than $5,000 per month per line dependent on the capacity, distance, and committed levels of services.

Cost savings gained through the use of a SIP system can include reductions in international and long distance (toll) usage charges and reductions in intelligent network feature costs. There may be no toll costs for calls terminated within a data network and the cost for terminating calls through gateways at remote locations (such as international calls) may be 50% to 90% less than traditional switched voice services. The cost of intelligent features such as the use of multiple local numbers instead of toll-free numbers can reduce the cost of intelligent network services.

Figure 8.3 shows some of the variable service costs that are associated with IP Telephony systems. This table shows that a company commonly pays a gateway access fee for calls terminating and possibly originating on remote gateways. The assignment of telephone numbers can cost $5 to $40 per assigned telephone number. For networks that connect through Internet data connections, the cost can range from zero (no change in committed capacity) to approximately $10 per user (increase in Internet data capacity). For systems that add leased lines or increase leased line capacity between concentrators (routers), the cost of leased lines can range from $40 to $5,000 per month per leased line.

Item	Cost
Gateway access fees	$0.01 to $0.05 per minute.
Telephone number assignment	$5 to $40 per month per number
Internet data connection	$0 to $10 per month
Leased Data Lines	$40 to $5,000 per month per line

Figure 8.3, SIP Service Costs

Other Costs

Other related costs for implementing SIP systems include installation, training support, consulting, software development, and taxes. Additional cost reductions come from integrated processes, simplification of service changes, and increased employee efficiency.

Installation can include running new data lines, adding routers to provide new data ports, and installing power hubs and backup power supplies to keep the SIP based telephone systems running when primary power is interrupted. Power to the IP telephones may be supplied locally by a power supply or it may be supplied from the data network through the use of power data hubs. Power data hubs use some of the unassigned data communication lines to supply power to devices that are connected to it. Power hubs are inserted between the data network (e.g. routers) and the IP telephones. Power hubs may be supplied with power from an uninterruptible power supply (UPS) to ensure the IP telephones operate when primary power is lost.

Training support includes end-user training and system administrator training. End- user training can be ½ to 1 hour of training by an internal staff member at a group session. The lost productivity of the worker time for training and the cost of the trainer and facility can result in a cost of approximately $15 to $30 per end-user.

Consultants may be hired to help evaluate system alternatives, data network design, SIP system design, and to oversee the installation, testing and integration of SIP systems. The cost of SIP system consultants can range from $85 to more than $250 per hour.

Custom software development projects may be issued to develop custom call processing features and to integrate SIP telephone systems with company information systems. Programmers within the company may perform software development or it may be outsourced to a software development company.

Taxes are fees imposed by government regulatory agencies for the selling of, or provision of, services. Some taxes may be collected directly from the end-customer and the service provider pays other taxes. Until recently, most governments did not apply taxes to IP telephony service providers. This preferred tax advantage is starting to change in many countries as regulators are starting to view IP Telephony providers as competitors to established telecom companies. The cost of taxes is not limited to the assessed fees; tax costs include the preparation costs of tariffs (for service providers) and tax reports.

Additional cost reductions that result from the use of SIP based systems include the use of standardized systems in multiple locations, reduced cost of 'Moves, Adds, and Changes' (MACs), and increases in worker productivity. The use of SIP-based systems allows the same technology to be used in multiple company locations. This allows for easy system interconnection and the establishment and management of a simple company-wide dialing plan. If existing data networks are used to provide telephony service, moves, ads, and changes of extensions may become an administrative function rather than requiring the installation of new cabling and wiring change work orders. The cost of worker operations will be reduced due to the advanced communication capabilities such as more advanced call and message distribution features.

Figure 8.4 shows some of the other indirect costs of installing and operating SIP systems. This table shows that some of the hidden costs include the cost of adding new data lines, training, consulting, custom software development, and taxes. This table shows that the data line installation costs may include running new lines (e.g. category 5 cable), adding routers for new data ports, and adding a power hub to supply power to IP telephones from a central location. Training costs include end-user training and system administrator training. Consultants typically charge $85 to $250 per hour to assist with the selection, installation, and testing of IP telephony systems. Software development costs for specialized features can range from $5,000 to more than $200,000 per project. The cost of taxes can be 1% to 10% of usage based fees.

Item	Cost
Installation	
- New line	$250 per line
- New data port	$50 per port
- Power Hub	$15 per port
Training	
- End User (1/2 hour to 1 hour)	$15 to $30 per user
- System Administrator (1 to 5 days)	$500 to $2,000
Consulting	$85 to $250 per hour
Software Development	$5,000 to $200,000+ per module
Taxes	1% to 10%

Figure 8.4, Other SIP Costs

References:

. Internet Telephony Expo, Grandstream, Long Beach, CA, October 15, 2003.
i. "FCC Study on Telephone Trends," Federal Communications Commission, May 22, 2002, 14-1.
. www.Linux.org
. www.Vovida.org

Chapter 9

How SIP is Changing

SIP and its associated protocols were designed to allow for changes that will permit new communication services and enhanced systems integration. To enhance the capabilities of SIP systems, the basic SIP protocol can be extended through the use of protocol extensions and SIP development toolkits. Some of the key extensions for SIP include SIP for Instant Messaging and Presence Leveraging Extensions (SIMPLE).

SIP and its associated protocols are relatively new standards. The core SIP protocol was intentionally designed to be simple in its design and to allow for future extensibility. It is through this extensibility that many of the new SIP services and features are appearing.

As SIP hardware appears in the market, so does a series of software tools aimed at product and application developers. There are now a range of SIP Toolkits available that allows SIP products, services and features to be readily created. Some of the SIP extensions that have been agreed recently or are in the process of formalization include presence and Instant Messaging (IM). Another recent extension of SIP is the Info method that can be used to transfer mid-session control information such as DTMF signaling tones.

Changes to SIP and supporting protocols are made by the Internet Engineering Task Force (IETF) SIP working group. Because the addition of new features and capabilities to SIP can potentially cause unexpected equipment operation results, the SIP working group extends the protocol when there is significant documented reasons to extend the protocol.

SIP Extensions

SIP Extensions are a set of commands or protocols that are used to extend the capabilities of the core SIP protocol. SIP extensions are commonly used to rapidly extend the capabilities of an existing application or protocol without changing the underlying application or protocol.

Figure 9.1 shows how SIP extensions can be used to enhance the capabilities of the SIP core protocol and how they interact with the existing core protocol. This diagram shows that SIP extensions are used to process new types of messages that are sent on the SIP system. When core messages are sent between the SIP devices (such as between proxy servers), they are processed for the function defined in the standard. If a new type of message is received, the SIP device will determine if the message is associated with a SIP extension that is has installed. This diagram shows that the message processed by the SIP extension may interact with the core protocol and other SIP extensions.

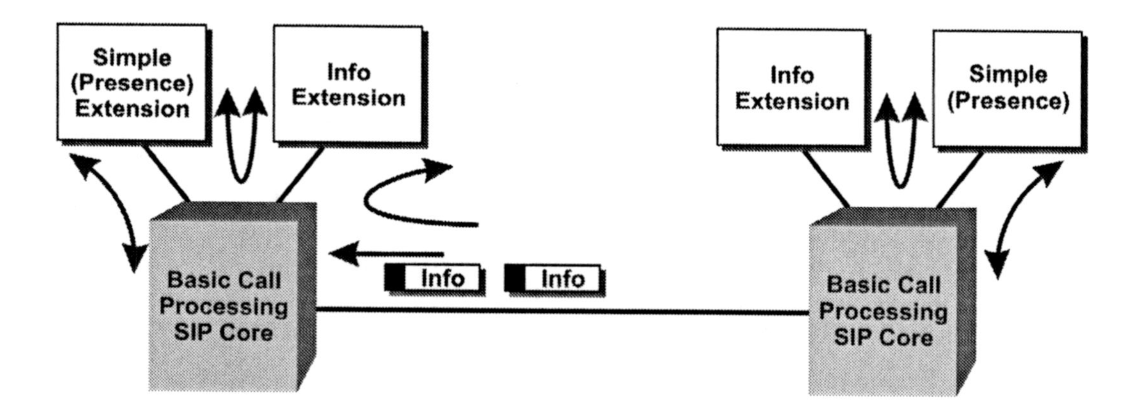

Figure 9.1., SIP Extensions

SIP Toolkits

SIP Toolkits are a group of software programs that assist designers or developers to create products or services that use the Session Initiation Protocol (SIP). These software tools will allow vendors to create products such as SIP end-points (IP telephones and messaging systems), SIP servers (proxy, redirection, and registrar), media gateways, and SIP application servers.

The toolkits are essentially SIP protocol stacks that support the core SIP specification, that is the ability to establish and mange multimedia sessions. Usually the toolkit will also support the extensions to the SIP protocol such as event notification, (which can be used to build presence services). The toolkit should be capable of operating on the most common hardware platforms under most operating systems, and will offer the developer a range of Application Programming Interfaces (APIs) through which to build services and features.

Call Processing Scripts and Servlets

A script or a Servlet is a small program or sequence of operations (macro) that is written in a predetermined language that can be understood by the calling program to allow automatic interaction between programs or devices. Examples of scripts include login scripts that are used to provide identification information when accessing a system or Javascripts that provide advanced features on Internet web pages.

Scripts usually have the ability (defined script commands) to receive, process, and return information (parameters) between the calling program and other programs and devices. This capability allows developers, programmers, and in some cases end users, to create advanced services that are not possible through the use of existing protocol standards (such as SIP).

Some of the more common methods for service creation in the SIP environment include the use of scripts such as SIP Common Gateway Interface (CGI), Call Processing Language (CPL), and SIP Servlets. SIP CGI is based on the Web CGI used to create interactive services at a web site; services can be created by loading CGI scripts on a SIP server. These scripts would be triggered under certain conditions and would influence call processing, for example to forward a call to a specified endpoint. This concept of triggering is very similar to the Intelligent Network (IN) concept found in traditional telephony networks, if a script is not triggered at a server the server will simply perform a default action.

When a request is received by a SIP server that is associated with a CGI script (a trigger), the CGI script program is started. The script program is initialized with the call information parameters so the service can be processed for that specific call or session. CGI scripts can interact with other SIP devices or other types of programs such as database queries, email systems, or voice mail systems. CGI scripts can modify messages as the pass through the proxy server extracting and changing message headers. The CGI script may be continuously provided with call status information from the proxy server during the call as the users may initiate actions, (such as selecting options or hanging up), that result in a change of CGI script operation. CGI scripts may also transfer and store information to and from other proxy servers in the form of tokens and cookies. This allows feature settings such as call processing preferences to be maintained for future sessions.

SIP standards do not specify the details of service operation. The messages and processes defined by the SIP standard are used with scripts and other information processing programs and tools to create advanced services. These scripts can be located in servers, user agents, or combinations of the two.

Figure 9.2 illustrates how a call screening service may be added to a SIP system through the use of a CGI script that is processed on a proxy server. In this example, a caller Bob dials Susan's phone number on a SIP phone. This creates an invite message that is received on the proxy server and the proxy server determines that this message requires the processing of a CGI script. This trigger starts the interaction between the proxy server and interactive voice response (IVR) unit. The script is used with the proxy server to set up an audio channel between the caller Bob and the IVR. This allows IVR to play a message that asks Bob to create an introduction message for Susan (the call screening process). When Bob has completed the message, the CGI script alerts Susan that she has an incoming call and if she answers, the CGI script through the proxy server connects Susan to the IVR so she can hear Bob's introduction. Assuming that Susan indicates she desires to answer the call (through keypad selection), the CGI script through the proxy server sets up an audio path between Bob and Susan.

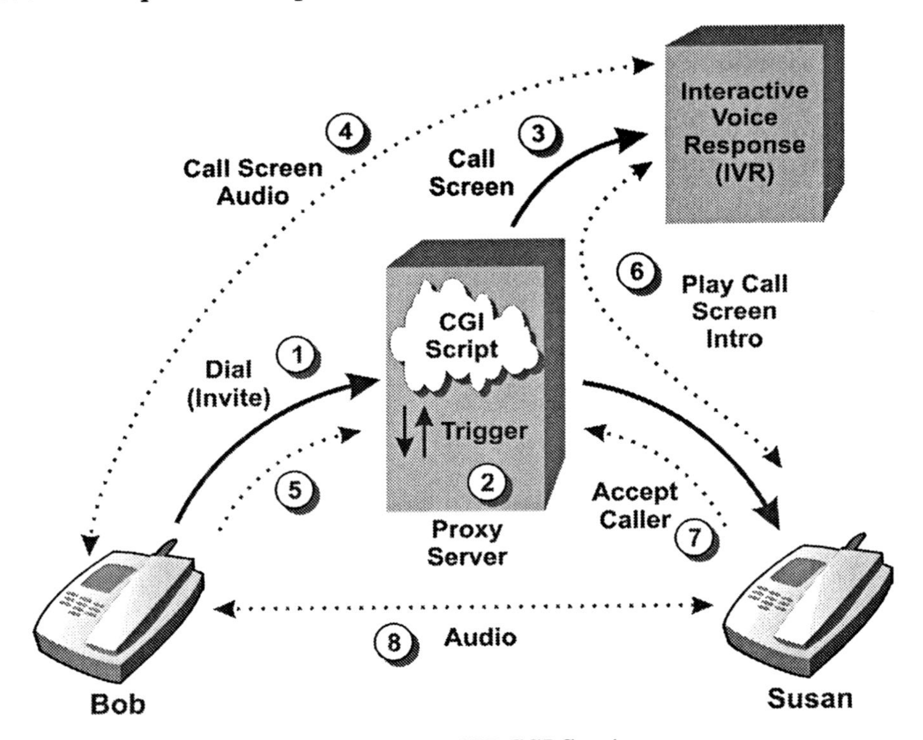

Figure 9.2., SIP CGI Service

SIP for Instant Messaging and Presence Leveraging Extensions (SIMPLE)

SIP for Instant Messaging and Presence Leveraging Extensions (SIMPLE) is an IETF working group that was formed in 2001 to address the numerous submissions to expand the capabilities of the session initiation protocol (SIP). A presence service allows a user to subscribe to presence information regarding others, so that the user may see when other users are connected to a network such as the Internet. By employing presence and messaging software users are able to build 'buddy lists' which indicate the current status of the people in the list. When another user is shown as available it is possible to use an Instant Messaging (IM) service to send and receive real-time messages.

Figure 9.3 shows the basic operation of SIP presence service. This example shows how an Instant Messaging (IM) user Pete desires to send an instant message to another IM user, Susan. This example shows that Pete and Susan are using a SIP based presence service that monitors outgoing messages (Presentity) and incoming messages (Watcher). This diagram shows that the presence service only manages the presence status of the users. The media (messages) are directly sent between the users.

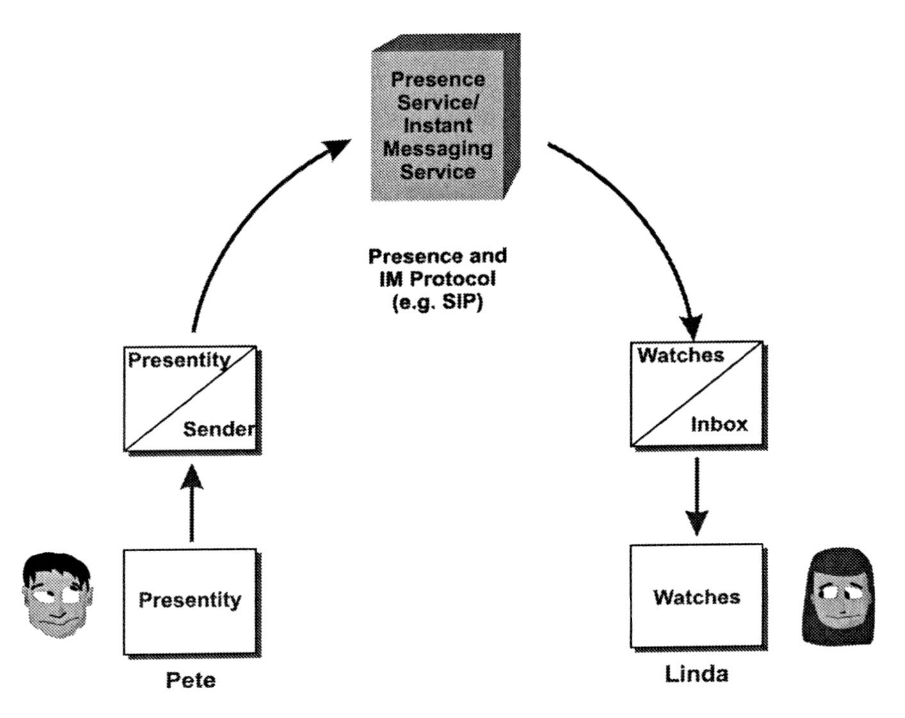

Figure 9.3., Generic Models for Presence and Messaging

In a presence service an entity known as a watcher subscribes for the service of the presentity of another user. After this time the watcher will be informed of the status of the other user by their presentity that generates notifications. At both ends of the association there is an appropriate User Agent that interfaces either to the presentity or the watcher as appropriate. Although some of the entities in a SIP based presence service may co-located, the model does not specify how these entities are distributed across network devices.

For Instant Messaging the watcher can now send a message to other users who are shown as available. The example shows the service in a unidirectional form, but of course by having identical entities at both ends, (both users with both a presentity and a watcher), and then presence information and messages could be sent in both directions.

Figure 9.4 shows how SIP extensions can be used to support Presence and Instant Messaging. This example shows how SIP is extended for Presence service with two new methods; Subscribe and Notify. In the diagram, the watcher (Larry) desires to know the status of his friend Susan show he subscribes to presence information at a presence service (server). This results in Larry receiving notifications of status changes relating to the presence of Susan. Once presence is established, the third SIP extension known as Message can be used to transfer Instant Messages from the watcher.

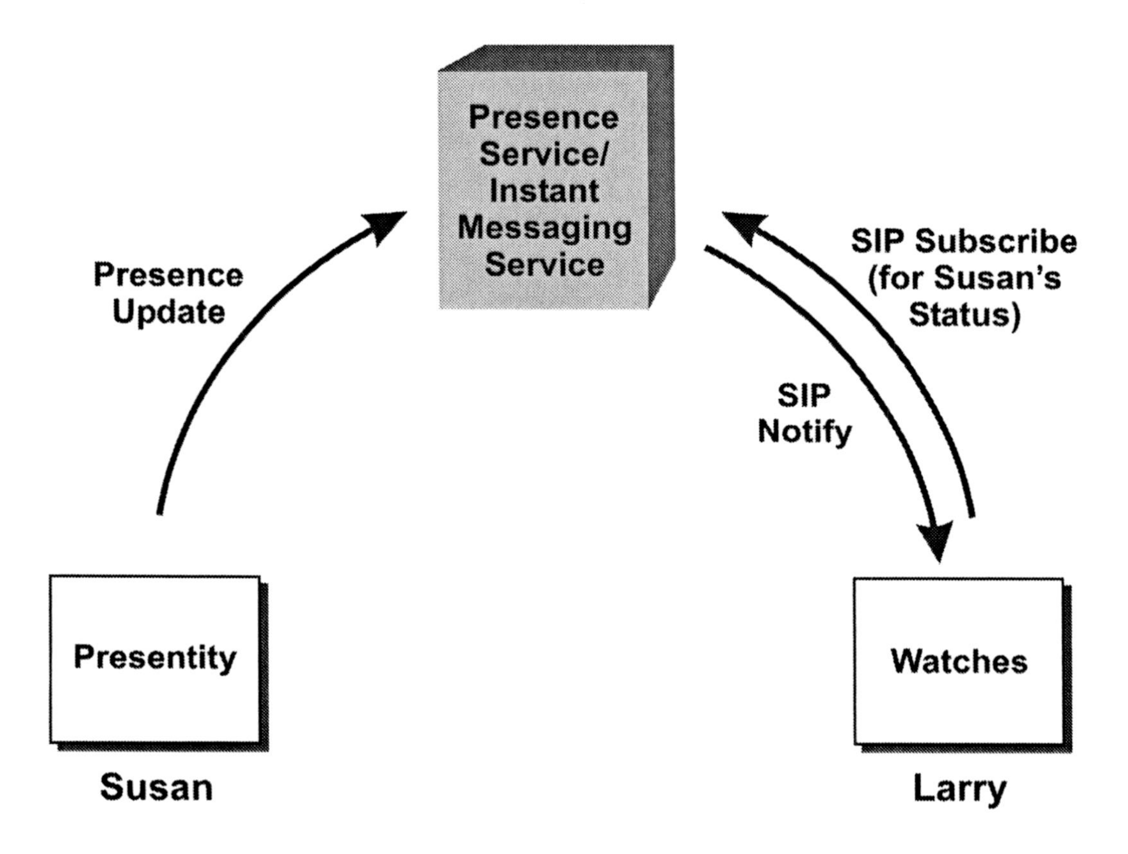

Figure 9.4., SIP Presence and Instant Messaging

Instant Messaging (IM) for presence service can occur either within a SIP session or can be outside any session. If two users have a SIP session established (such as during a telephony call), an IM could be transferred within the established session. For IM service between users that do not have an established session, each message is treated as a unique entity.

The SIP Subscribe and Notify methods were not written specifically with Presence services in mind. SIP Subscribe and Notify methods can be used to construct other services such as call back and message-waiting indication associated with voicemail systems. This demonstrates the power of the SIP protocol that is written around a limited set of simple methods on which a range of complex services can be built.

SIP Info Method Extensions

SIP Info method extensions are used to provide control information that relates to the communication session. SIP Info can be used to convey mid-session information that relates to the application layer. For example, this information could be DTMF tones generated from a telephone keypad during a call. DTMF tones may be used to allow users to interact with an automated system (for example to make menu selections or for entering account information). The Info method could also be used to convey non-streaming information between the users in a session. This includes insertion of new media such as a picture that is transferred between the users. The Info method does not change the state of the call, it is simply a technique to transparently convey application level information along the SIP signaling path.

Figure 9.5 illustrates how the SIP Info extension can be used to convey DTMF tones to an interactive server. This example shows a user who is calling to a SIP based system and is accessing an interactive service. The interactive server provides voice prompts to the caller to allow the caller to make various menu selections using their keypad. The resulting DTMF tones are converted to SIP Info format and transferred to the server.

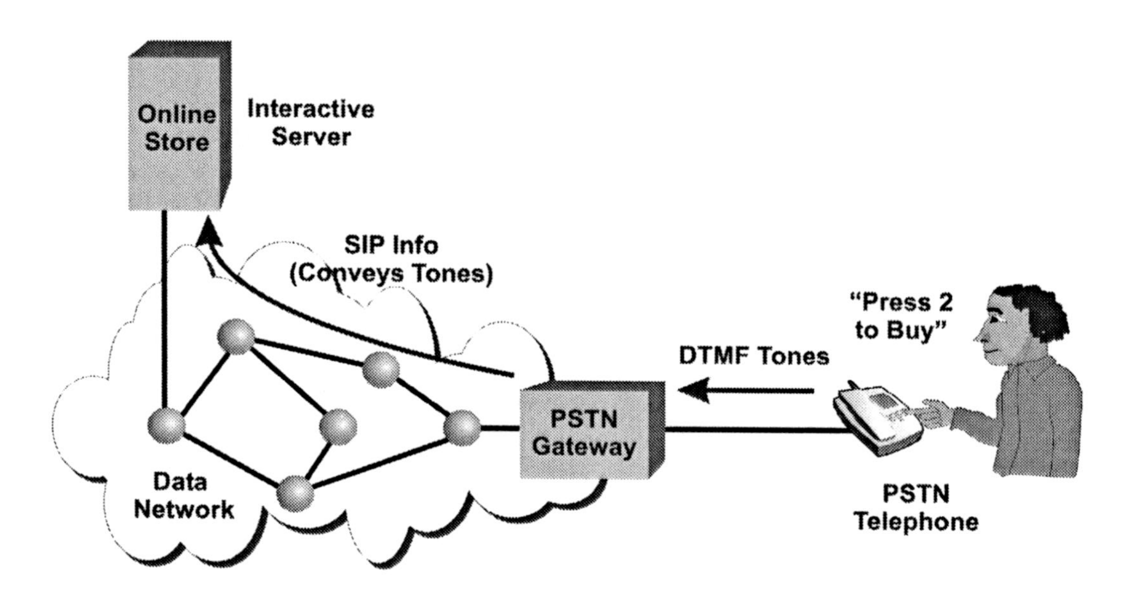

Figure 9.5., The SIP Info Method

Index

About Our Instructors

Mr. Harte is the president of Althos, an expert information provider covering the communications industry. He has over 29 years of technology analysis, development, implementation, and business management experience. Mr. Harte has worked for leading companies including Ericsson/General Electric, Audiovox/Toshiba and Westinghouse and consulted for hundreds of other companies. Mr. Harte continually researches, analyzes, and tests new communication technologies, applications, and services. He has authored over 30 books on telecommunications technologies on topics including Wireless Mobile, Data Communications, VoIP, Broadband, Prepaid Services, and Communications Billing. Mr. Harte holds many degrees and certificates include an Executive MBA from Wake Forest University (1995) and a BSET from the University of the State of New York, (1990).

Mr. Bowler is an independent telecommunications training consultant. He has almost 20 years experience in designing and delivering training in the areas of wireless networks and related technologies, including CDMA, TDMA, GSM and 3G systems. He also has expertise in Wireless Local Loop and microwave radio systems and has designed and delivered a range of training courses on SS7 and other network signaling protocols. Mr. Bowler has worked for a number of telecommunications operators including Cable and Wireless and Mercury Communications and also for Wray Castle a telecommunications training company where he was responsible for the design of training programmes for delivery on a global basis. Mr. Bowler was educated in the United Kingdom and holds a series of specialized maritime electronic engineering certificates.

Typical Training Costs

On-Site Training

Althos on-site training cost ranges from $2,200 to $3,600 per day of instruction plus expenses dependent on the length of the course and the type of content (labs, exercise materials, and instructor skill level). Althos does not charge for instructor travel time.

If Althos must incur travel expenses in conjunction with the project (this is typical for an on-site training session or presentation), travel expenses will be reimbursed on the basis of actual cost. The client prior to commitment will approve all travel expenses.

Online Training (Web Seminars)

Althos provides some courses and executive briefings in the form of online web seminars. Althos web training seminars allow two-way audio with all the participants along with presentation materials. The typical cost of web seminars range from about $85 for a 1-hour open enrollment executive briefing to approximately $350 per day for standard course instruction.

Open Enrollment

Althos periodically offers open enrollment to allow individuals to attend courses. The typical cost for individual enrollment ranges from $1,100 to $1,800 per student dependent on the location and type of course. Open enrollment courses include meals and materials (books and workbooks).

Custom Course Development

Althos can customize our courses to meet your specific training need or we can research and use your materials to create a new course. Custom course development fees range from $50 to $200 per presentation slide (graphics + descriptive text).

Printed in the United States
1501200001B/71